PALGRAVE STUDIES IN THE HISTORY OF
SCIENCE AND TECHNOLOGY

James Rodger Fleming (Colby College) and Roger D. Launius (National Air and Space Museum), Series Editors

This series presents original, high-quality, and accessible works at the cutting edge of scholarship within the history of science and technology. Books in the series aim to disseminate new knowledge and new perspectives about the history of science and technology, enhance and extend education, foster public understanding, and enrich cultural life. Collectively, these books will break down conventional lines of demarcation by incorporating historical perspectives into issues of current and ongoing concern, offering international and global perspectives on a variety of issues, and bridging the gap between historians and practicing scientists. In this way they advance scholarly conversation within and across traditional disciplines but also to help define new areas of intellectual endeavor.

Published by Palgrave Macmillan:

Continental Defense in the Eisenhower Era: Nuclear Antiaircraft Arms and the Cold War
By Christopher J. Bright

Confronting the Climate: British Airs and the Making of Environmental Medicine
By Vladimir Jankovic´

Globalizing Polar Science: Reconsidering the International Polar and Geophysical Years
Edited by Roger D. Launius, James Rodger Fleming, and David H. DeVorkin

Eugenics and the Nature-Nurture Debate in the Twentieth Century
By Aaron Gillette

John F. Kennedy and the Race to the Moon
By John M. Logsdon

A Vision of Modern Science: John Tyndall and the Role of the Scientist in Victorian Culture
By Ursula DeYoung

Searching for Sasquatch: Crackpots, Eggheads, and Cryptozoology
By Brian Regal

Inventing the American Astronaut
By Matthew H. Hersch

The Nuclear Age in Popular Media: A Transnational History
Edited by Dick van Lente

Exploring the Solar System: The History and Science of Planetary Exploration
Edited by Roger D. Launius

The Sociable Sciences: Darwin and His Contemporaries in Chile
By Patience A. Schell

The First Atomic Age: Scientists, Radiations, and the American Public, 1895–1945
By Matthew Lavine

NASA in the World: Fifty Years of International Collaboration in Space
By John Krige, Angelina Long Callahan, and Ashok Maharaj

Empire and Science in the Making: Dutch Colonial Scholarship in Comparative Global Perspective
Edited by Peter Boomgaard

Anglo-American Connections in Japanese Chemistry: The Lab as Contact Zone
By Yoshiyuki Kikuchi

Eismitte in the Scientific Imagination: Knowledge and Politics at the Center of Greenland
By Janet Martin-Nielsen

Climate, Science, and Colonization: Histories from Australia and New Zealand
Edited by James Beattie, Emily O'Gorman, and Matthew Henry

The Surveillance Imperative: Geosciences during the Cold War and Beyond
Edited by Simone Turchetti and Peder Roberts

Post-Industrial Landscape Scars
By Anna Storm

Voices of the Soviet Space Program: Cosmonauts, Soldiers, and Engineers Who Took the USSR into Space
By Slava Gerovitch

After Apollo? Richard Nixon and the American Space Program
By John M. Logsdon

Frontiers for the American Century: Outer Space, Antarctica, and Cold War Nationalism
By James Spiller

Frontiers for the American Century

Outer Space, Antarctica, and Cold War Nationalism

James Spiller

FRONTIERS FOR THE AMERICAN CENTURY
Copyright © James Spiller, 2015.

First published in 2015 by
PALGRAVE MACMILLAN®
in the United States—a division of St. Martin's Press LLC,
175 Fifth Avenue, New York, NY 10010.

Where this book is distributed in the UK, Europe and the rest of the world,
this is by Palgrave Macmillan, a division of Macmillan Publishers Limited,
registered in England, company number 785998, of Houndmills,
Basingstoke, Hampshire RG21 6XS.

Palgrave Macmillan is the global academic imprint of the above companies
and has companies and representatives throughout the world.

Palgrave® and Macmillan® are registered trademarks in the United States,
the United Kingdom, Europe and other countries.

ISBN: 978–1–137–50786–0

Library of Congress Cataloging-in-Publication Data is available from
the Library of Congress.

A catalogue record of the book is available from the British Library.

Design by Newgen Knowledge Works (P) Ltd., Chennai, India.

First edition: October 2015

10 9 8 7 6 5 4 3 2 1

For A., A., A., & T.

Beloved partners in the expedition of life

In memory of M.

Contents

Figures

Acknowledgments

Scholars often specialize in fields that align with their personal interests. So it is not surprising that I leaned for a time toward the history of United States exploration in outer space and Antarctica. In addition to a childhood of punctuated awe over moon landings and early planetary missions, adolescent memories include looking out over New Jersey cliffs as my bonfire friends pretended overhead planes were starships. Those happy nighttime romps turned to daytime rock climbing, which took me far from New York City suburbs to western wilds where I honed my skills as a backcountry explorer. As I joined and then led wilderness mountaineering expeditions, I dreamed of more remote and rugged destinations. Epic tales of Antarctic exploration piqued my interest in what was then an incomprehensively distant locale.

The academic alignment with my passion for exploration and expeditions emerged during my undergraduate senior thesis, my first foray in the history of American culture and spaceflight. I remain indebted to Paul Israel, James Livingston, and the late Philip Pauly, historians who advised that rough work and saw in it the potential of a future scholar. With their encouragement, I continued my studies at the University of Wisconsin where I had the great fortune of learning from the late cultural historian and master editor Paul Boyer, the prolific and penetrating historian of science Ronald Numbers, and the consummate academic mentor and environmental historian William Cronon. Thanks to them, the sagely geographer Yi Fu Tuan, and such supportive colleagues as Andrew Rieser and Susan Traverso—the first to encourage my interest in the southern continent— I finished graduate school with a first round comparative study of the cultural politics of US space and Antarctic exploration.

My research during and since my graduate studies was made possible by the capable staff of: Wisconsin Historical Society; National Archives; Library of Congress Manuscript and Film & Television Archives; Smithsonian National Air and Space Museum (NASM) Archives; NASA History Office; Hagley Library; Byrd Polar and Climate Research Center

Archives; Lyndon B. Johnson Presidential Library; Gerald R. Ford Presidential Library; Jimmy Carter Presidential Library; George H.W. Bush Presidential Library; Naval Historical Center; and the Interlibrary Loan desk at The College at Brockport, SUNY. I appreciate their assistance and critical work to steward our heritage, the evidentiary bases of scholarship and civic vitality. I bear a special debt to the scholars in the Space History Division of NASM. Robert Smith, then head of that division, was a valued sponsor, and current NASM Associate Director Roger Launius has long been a model and indispensable resource as the dean of space history.

I hope my modest scholarship does some justice to their good works and warrants the patient indulgence of my Brockport History colleagues, whose commitment to matching first-rate scholarship with outstanding teaching has kept this restless explorer committed to the academy. That commitment has broadened through the administrative opportunities afforded by Anne Huot and Timothy Killeen, visionary educational leaders and now university presidents who generously included me in their projects to sustain academic research.

That project is critical due to the fast-changing budget model of research and higher education. What remains true for many researchers was the case for me; my work was made possible by fellowships and grants. For their generous financial support, special thanks to: US Department of Education; Smithsonian Institution; University of Wisconsin Foundation; President Lyndon B. Johnson Foundation; Gerald R. Ford Presidential Foundation; George H. W. Bush Presidential Library Foundation; and The College at Brockport faculty grant program.

Special thanks also go to Jonathan Cherin and my backcountry partners Gary Gordon and Bruce Nilles. These friends and climbing partners have held fast my many ropes and avoided indelicate talk of long delayed manuscripts.

Last, I hope this work is worthy of dedication to: Arlene, for life itself and providing the ethical compass for sure navigation; Anne, a beloved partner who indulges my many tacks and a talented scholar and teacher who presses for good follow through; and Aelis and Tamsin, for manifesting beauty and verve and giving transcendent meaning to this group's journey together.

Introduction: Polar *Stars and* Stellar *Stripes*

A half century ago more than fifty million people glimpsed a familiar chrome-plated vision of the future at the 1964 New York World's Fair (Figures 0.1 and 0.2). Its many dioramas of a glittering world to come cast American science and technology in leading roles, driving the fast moving civilization on display. They reminded visitors that science had unleashed human genius while technology improved their lot, delivering them from the drafty hovels of the past to the modern comforts of an electrified landscape of glass, steel, and concrete. None did so with greater flare than Futurama II, the pavilion sponsored by auto giant General Motors. The exhibit depicted American know-how and machinery bearing untold bounty to 21st Century "Man" as he systematically harvested the oceans, turned arid wastes into irrigated cropland, and cleared fetid jungles for productive farms and industry. More fantastic still, it visualized a new epoch in Man's command over his surroundings in which the farthest reaches of the world and beyond— Antarctica and outer space— finally yielded to human control. GM's miniature display of atomic powered bases at the South Pole and on the moon flew tiny Stars and Stripes, clearly flagging them as manned by the United States. Futurama II channeled the Fair's claim to be "Universal and International," portraying them as frontier outposts of a global civilization. The domed bases poignantly signified that America used its matchless science and technology to conquer new frontiers for all humankind.[1]

The gleaming certainties of GM's exhibit captured the zeitgeist of a nation, what one historian calls Americans' "grand expectations" that they faced "no limits to progress."[2] Those limits became painfully evident only a few years later. But the 1964 New York World's Fair expressed the still common optimism that America's wondrous advance would continue, for it seemed to have history's favorable winds in its sails. The nation's atomic age leadership of a growing

Figure 0.1 Depiction of a US Antarctic station in the General Motors Futurama II exhibit at the 1964 New York World's Fair.

Figure 0.2 Depiction of an American lunar base in the General Motors Futurama II exhibit at the 1964 New York World's Fair.

"Free World," coming quickly after it pulled itself out of an industrial depression and scored a technology-intensive victory during World War II, sensibly came off as the prelude to an even more dazzling space age. The vertiginous dawn of this new era triggered an explosion of science and technology that energized the American economy and touched the lives of people everywhere. Fortified with fast mounting knowledge and tools to harness the awesome power of the atom and turn the numinous firmament into an arena of routine activity, Americans reasonably harbored grand expectations that they would positively remake their world. They would even assume dominion over the ends of the earth, finally making Antarctica and outer space do their bidding and serve the rising aspirations of humankind.

These grand expectations were not simply spontaneous effects of a fortuitous run of national power and prosperity. National leaders who fervently believed in America's bright future and righteous global leadership had assiduously cultivated those expectations for more than 20 years. They understood this rosy future depended on the support of the American people, many of whom warily regarded atomic age science and technology and previously opposed international entanglements. So public leaders tirelessly declared the United States the necessary head of the "Free World" and promised that its benevolent leadership and wellspring of research and innovation would deliver a more perfect world for all humankind. Thus the movers behind the New York World's Fair did not simply channel the uncontrived sentiments of a forward looking nation. They captured the hegemonic sensibilities of a society resounding with public celebration of the "American Century," a common label for a dawning age of peace and prosperity made possible by America's benevolent world leadership. That leadership, in turn, was made possible by the country's cutting-edge science and technology.

The people who designed Futurama II also tapped a more specific supposition in wide circulation, namely that America's benevolent power and global purpose were strikingly evident at the ends of the earth. Public officials and the reportorial class had advanced this idea since the 1957–58 International Geophysical Year (IGY), when more than 60 nations across the Cold War divide came together to study the whole of the earth and its oceans and atmosphere. During that year the United States sent researchers to Antarctica and planned to orbit the world's first satellite. It did so ostensibly for the high minded interests of science, so as to feed measurements from these hard to reach places into the IGY geophysics databank. Not to be outdone, the Soviet Union did the same. The geopolitical import of these IGY programs became evident when the Soviet Union impressed the world with its scientific and technological prowess by beating the United States into space in October 1957. Many Americans were shocked and felt diminished by the Soviet Sputnik, humankind's first artificial satellite.[3] In response, national leaders attempted to assert US leadership in these fields by drafting science advisors for the President and boosting math, science, and engineering education with federal grants. The United States also turned its temporary IGY initiatives in Antarctica and outer space into permanent government programs. The National Science Foundation (NSF) and the new National Aeronautics and Space Administration (NASA) managed these respective programs with an eye to assisting the United States in its Cold War struggle for international prestige. After Sputnik these programs became conspicuous testaments to the country's preeminent science and technology and lofty aim to secure an American Century, during which the United States would share the benefits of its growing dominion over earth and beyond with all people. After several years of such striking testimony, fairgoers were probably not surprised to see scale models of US stations in Antarctica and outer space on display in Futurama II. These miniature outposts perfectly illustrated the Fair's grandiose theme, "Man's Achievement on a Shrinking Globe and in an Expanding Universe," and they exemplified the common assumption that these untamed regions had become humankind's final quarry. [4] They had become frontiers for the American Century.

The American Century and Cold War Nationalism

The nationalist myth of the frontier enjoyed enormous currency at mid-century and captured well the enduring axiom of American exceptionalism.[5] Even before the nation's founding, European settlers often described their New World experiment as a cut above other societies. It was a proverbial city on the hill to inspire spiritual uplift and secular progress in a wayward world.[6] When the United States became a great world power in the late nineteenth century, historian Frederick Jackson Turner gave voice to common intellectual stirrings by updating this story of national exceptionalism. He determined that Americans had enjoyed a unique mix of freedom, democracy, and prosperity afforded by their continental frontiers. Turner saw America's singular virtue as threatened, owing to the closure of these frontiers, and at best as an inspiration to other nations. His sympathetic but more global minded contemporary Theodore Roosevelt was more optimistic. He too saw the frontier as the source of America's virtue. However the future president promoted the spirit of the frontier as an ongoing fount of national vitality and urged his countrymen to cultivate the energy of early backwoods explorers so as to vigorously assert their influence around the world. Americans would thereby sustain their own wealth and pioneering spirit while pushing others toward open markets and societal progress, which he considered an enlightened alternative to Europe's imperialist offerings to the world. America's frontier character was the bridge between its hallowed past and Roosevelt's preferred future. It made an uncharted break into international affairs appear comfortably consistent with the nation's historic trajectory.

Decades later after the country's incipient internationalism foundered during the Great Depression and early years of World War II, many high placed Americans determined that only US leadership could save modern civilization from economic collapse and its crazed descent into war. Prominent among them was Henry Luce, founder of the enormously popular *Time* and *Life* magazines. In his February 1941 *Life* essay "The American Century," Luce implored Americans to recognize that the United States could no longer remain just a beacon to the world. The country had become the world's preeminent power, he counseled, and now had to assume providential leadership of this American Century and save the world from fascist conquest. Luce admitted the United States could remain aloof as "the entire rest of the world came under organized domination of evil tyrants." But he believed "Peace cannot endure unless it prevails over a very large part of the world," and he offered US internationalism as the path to this better alternative. Only America could defeat the Axis menace and bring true progress to the world "in terms of a vital international economy and in terms of an international moral order." Since his geopolitical vision alone was unlikely to galvanize his isolationist peers, Luce appealed to their nationalist sensibilities as well. Like Theodore Roosevelt, he argued that global leadership was essential to America's wellbeing *and* that it was true to the nation's history and noble character. Luce accordingly painted a picture of the American Century that entailed "a sharing with all people of our Bill of Rights, our Declaration of Independence, our Constitution, our magnificent industrial products, our technical skills" so as to secure a "more abundant life" for all "predicated on Freedom."[7]

Luce hoped to inspire Americans, in historian Alan Brinkley's words, "to undertake a great mission on behalf of what he considered the nation's core values," namely "to play a forceful role in both ending [the war] and building a better world in its aftermath." Although his vision then enjoyed "relatively little broad public support," Americans wholeheartedly took up the first part of that mission in December 1941 when they went to war after the Japanese attacked Pearl Harbor, Hawaii. Public support for the follow-up mission, building a better peacetime world, was less clear even as many leaders determined that the United States had to build and safeguard a liberal international order after the war. Vice President Henry Wallace, for instance, championed US leadership of such a postwar order. His 1942 call for a postwar "Century of the Common Man" was largely consonant with Luce's American Century and with a good deal of wartime government and corporate propaganda. These propagandists worked to buoy public support for wartime mobilization and postwar global leadership by describing a world divided between fascist slave masters and freedom loving people, many of whom desperately fought alongside the United States. If America continued standing with them after the war, many internationalists averred, it could turn a murderous century into a peaceful and prosperous one. All freedom loving people could enjoy "a better standard of living," Wallace declared, "not merely in the United States and England, but also in India, Russia, China, and Latin America."[8]

Americans who previously cared little for the standard of living of these distant people remained cool to this evangelical internationalism. They responded more favorably during and after the war to stark warnings that the United States could no longer escape a global battle between freedom and tyranny. The Japanese had pulled the United States into that terrible battle, a new age of annihilationist warfare in which authoritarian despots proved willing to destroy whole cities and wipe out whole peoples. Even though Americans generally saw their nation as innocent of this cruel savagery, many feared that the atomic bomb the US developed and used during the war made an uncertain world all the more vulnerable to the hellish designs of totalitarian regimes. That fear galvanized American support for energetic international leadership shortly thereafter, when wartime amity degenerated into a Cold War between the United States and the Soviet Union. As Soviet troops occupied much of postwar Eastern Europe, that communist country quickly went from wartime ally to postwar foe. The apparent successor to the Axis menace, it seemed a totalitarian empire bent on world conquest. This menace loomed even larger in American minds by 1950. Russia had exploded an atomic bomb, communist revolutionaries won control of China, and Moscow's warmongering ally in Pyongyang tried to do the same over the Korean peninsula. Widespread fears of expansionist communists now armed with atomic weapons strengthened public support for US global leadership, so that it became not only the principal of international diplomacy and trade but also the military power needed to contain communism around the world.

US leadership of the Free World lasted for more than 40 years of Cold War and entailed far greater costs than Henry Luce foresaw when he called forth an American Century. Whereas Luce vaguely imagined a battered world reflexively embracing America's principled leadership, the country's international authority rested in fact on the sometimes costly and coercive application of its hard economic and military power. The US economy towered over those of the rest of the world, and its armed forces were simply overwhelming. Although many shattered countries and newly independent ones welcomed American support or had little choice but to work with the United States to protect themselves and rebuild their economies, the Cold War was a hegemonic project for the United States. It could not sustainably exert international leadership without the abiding support of people at home and abroad.[9] American internationalists consistently highlighted their country's benevolent aims, encouraging these people to believe that America's global leadership was just and that it would rebound to their benefit.[10]

The United States was similar in that respect to its Cold War adversary. They were unlike other world powers that had justified their colonial takings as benevolent paternalism. Instead, America and the Soviet Union each vied for international leadership by offering the fruits of its hard power and the allure of its utopian ideology. These Cold War superpowers promised partnerships that would shield their global brethren from the depredations of the other and put them on the fast track to a prosperous modernization. Soviet propagandists claimed that communism could do so because it was the most just and productive form of human affairs. They offered as proof their country's rapid transformation from feudal impoverishment to industrial world power. The Soviet Union was ready to share the prodigious achievements of its proletarian revolution with people elsewhere, enabling them to quickly achieve the intellectual and material development, Marxist doctrine held, they universally desired and equally deserved. Their American counterparts insisted that liberty, pluralism, and free enterprise were in fact the true measures of human nature, the best ways to unleash innate talents of people everywhere so that they could rapidly improve themselves and advance their societies.[11] These exponents pointed to the even greater bounty of America's production as evidence of the natural superiority of its brand of democracy and capitalism, which Henry Luce called "the ideals of civilization."[12] Luce's idealistic vision of American power put into service for human civilization applied well to this global battle for hearts and minds, and many people used his catchphrase thereafter to invoke America's mighty power and "its crusading internationalism during the Cold War." The American Century endured "as a description of America's continuing image of itself," what Alan Brinkley calls an image of a nation "that sets the course of the world's history— a nation whose values and virtues continue to make it a model to other peoples."[13]

Champions of the American Century drafted this image repeatedly on the Cold War homefront to reinforce a collective national identity suited to this crusading internationalism. They invoked a nation whose special character made it the best hope for human progress in a world dogged by enemies of freedom. Their national imaginings often came in sweeping profiles of America, such as those US presidents offered in rousing inaugural addresses when they spoke of their great nation's scared obligation for world leadership. These widely broadcast speeches were important distillations of national character and purpose. But this cultural project of national identification was generally more prosaic. Cold War internationalists repeatedly revealed those ideals and obligations through ordinary aspects of American life, which they claimed could positively affect the course of world history. Thus the US Department of Agriculture celebrated family farms as the essence of a free nation's pluck and productivity, the source of foodstuff and farming practices for desperately hungry people around the world.[14] Corporate boosters pointed to industrial harmony and worker compensation in their own firms as proof that America's free enterprise system engendered widespread prosperity at home and could provide the goods and manufacturing practices people everywhere needed to improve their lot.[15] The US Information Agency borrowed this script and treated middle-class homeowners and the expressways they drove to work as concrete signs of the speedy road to progress available to all who adopted America's political and economic systems.[16]

US leadership of the Free World depended on a national identity oriented to internationalism with the power to move Americans. It required an affective nationality that was invoked repeatedly in private forums and public venues and that was grounded in people's everyday experiences. The course of world history seemed open in this American Century to the salutary influence of those everyday experiences. As the 1964 New York World's Fair indicated, the fate of humankind seemed set above all else by a positively life changing feature of daily affairs, the awesome forces of American science and technology. These forces had dramatically improved Americans' health and standard of living, and they so clearly determined the balance of world power. If the USSR relied on its vast armies of conscripts, America's strategic advantage came from its high-tech weapons, especially the atomic and then thermonuclear bombs issued first by US research and development laboratories. Although the Soviet Union subsequently deployed these weapons too, the United States appeared to have an insurmountable advantage in science and technology that underscored its international leadership. Americans vowed to use that advantage to shield people from all out war and to improve their lives by sharing with them the many path-breaking innovations featured at the World's Fair. The promise of the American Century rested on these good intentions and on the obvious scientific and technological leadership of the United States, which Americans confidently regarded as their nation's birthright.

They were cocksure that science and the technology it spawned substantially improved the world and flourished best in the United States. As the leading light of US science policy Vannevar Bush explained, technological innovation and progress in all aspects of society "depends upon a flow of new scientific knowledge." That flow was strongest in the United States, he felt, owing to federal support for R&D and to the nation's legacy of freedom and democracy. Bush associated that legacy with America's frontier past, and his famous 1945 report *Science: The Endless Frontier* suggested that legacy could continue since "the pioneer spirit is still vigorous within this nation [and] Science offers a largely unexplored hinterland for the pioneer who has the tools for his task."[17] Bush used the familiar frame of American frontier nationality to express what historian David Hollinger called the widespread belief that "science and democracy were expressions of each other."[18] According to that common belief, science could only thrive where researchers freely pursued their individual genius and democratically vetted their findings with one another. As the National Science Foundation chief put it, science is inherently "democratic [and] has always been, in this sense, a 'free enterprise' system."[19] When the Soviet Union tested this liberal orthodoxy by developing an atomic bomb, Americans reasonably assumed that it had done so by pilfering their atomic secrets. In this vein, Vannevar Bush recognized that "Russia is a closely controlled dictatorship" that could "in the short run" drive its people to such success, especially when they copied the work of free people. But he remained calmly assured that in the "long run, a totalitarian state cannot compete with a free people in the advancement of science, for the dictation and dogma are contrary to the free spirit of inquiry, which is the heart's blood of scientific advance."[20]

Such calm assurance failed when Soviet scientists and engineers leapt ahead of their Free World counterparts and put the world's first satellites into earth orbit. The ballistic missiles that launched these Sputniks undercut America's strategic military advantages and put into question a key premise of the American Century, that only the United States could debut the path-breaking weapons and life-enhancing science and technology befitting the world leader. *The New York Times* worried that people might mistakenly believe "that Moscow has taken over world leadership in science," and Henry Luce's *Life* magazine warned that the "Sputniks give this old Communist swindle a new lease of plausibility."[21] National leaders quickly agreed they had to put this swindle to rest by besting the Soviet Union in outer space. As a ranking government official explained, a leading US space program would "be construed by other nations as dramatically symbolizing national capabilities and effectiveness."[22] Such a program would generate "a public image of supremacy," a US congressman declared, and impress "the peoples of the world of that reality of [American] power."[23] The prominent science advisor Lloyd Berkner agreed. He knew well that the American Century depended not only on the reality of that power but also on people's *belief* that the United States remained preeminent in science and technology. The domestic morale and national prestige necessary for world leadership were at stake since cutting-edge R&D had plainly become critical to the global balance of economic and military power. Each superpower jockeyed to maintain that edge and curry people's belief that it had done so since scientific and technological leadership signified its ideological superiority, military preparedness, and ability to enrich itself and its allies.

Lloyd Berkner was an architect of the IGY and knew that the United States had already worked to bolster that belief through its satellite and Antarctic initiatives. On the advice of men like Berkner, the White House and Congress turned these IGY efforts into permanent programs after Sputnik. National security now depended on reconnaissance satellites and on a strong enough presence in Antarctica to contain the regional designs of an emboldened Soviet Union. While the US space and Antarctic programs secretly bore these geopolitical aims, NASA and the National Science Foundation openly managed them to enhance America's stature by demonstrating its scientific and technological leadership. Agency spokesmen and media publicists appointed the United States a scientific trail blazer who used these high-tech programs of exploration to make the icy Antarctic and the boundless cosmos serve humankind. Taking a page from Vannevar Bush, who pushed science as America's endless frontier, they often depicted these costly national programs through the salient frame of American frontier nationality. They described a pioneering nation ready to demonstrate its vim and vigor once again and support a free and prosperous global civilization by conquering these final frontiers. Since US leadership of the Free World was predicated on such exceptional prowess and dedicated purpose, outer space and Antarctica became perfect frontiers for the American Century. As Lloyd Berkner explained, "In our day, when few physical frontiers remain, peoples visualize space and the Antarctic as challenges that must be accepted by a great nation to demonstrate its mettle."[24]

Berkner and other boosters who were professionally or politically invested in these costly federal programs solicited public support for them by pushing the trope of the space and Antarctic frontiers. They did so not simply for personal gain. They believed these frontiers were ripe for the taking and that such an effort was a worthy task for America's powerful science and engineering networks, what one historian called its "military-industrial-academic complex."[25] It would confirm what supporters of that security-oriented complex often identified as its higher purpose, that it was in fact a mighty force for human progress. The keepers of public opinion thought so as well, and the motif of the space and Antarctic frontiers echoed among the many channels of public discourse. Just as it had in Roosevelt's time, the frontier motif made unprecedented turns in public affairs potentially more palatable to a wary public. The motif helped make the sudden growth of that technocratic complex and the country's new internationalist commitments to an American Century appear comfortably consistent with the nation's pioneering character and frontier history.

The motif of the space and Antarctic frontiers was thus *politically useful*. It became *politically effective* because it resonated with a two-stranded nationalist myth at the height of its popularity. The story of national progress and liberal vitality that Frederick Jackson Turner told captured the optimistic spirit of mid-century America, which resounded with confident talk that the pioneering United States would bring forth ever-greater freedom and prosperity. So too did the national storyline favored by Theodore Roosevelt, who hoped his moralistic chronicles of societal rejuvenation on America's frontiers would inspire his enervated peers to dedicate themselves to grand national endeavors. Sputnik unleashed latent anxieties that Americans had once again become enervated by materialism and self-absorption, which many pundits fingered as the sorry reason why they had fallen behind their spacefaring communist adversaries. Many repeated Roosevelt's solution to this dire problem and called on Americans to follow their frontier predecessors by bucking up and consecrating themselves to a higher cause. Conquering the space and Antarctic frontiers was certainly a noble cause that could fire up the nation and revive its dissipated ranks. Doing so also raised the Turnerian promise of abiding liberty and prosperity. The space and Antarctic frontiers demanded a supreme pitch of national performance that only free people could sustain, and these forbidding realms would reward Americans and their global brethren with boundless resources and economic opportunities.

The cosmos and the icy bottom of the world became frontiers for the American Century because the space and Antarctic programs fulfilled basic geopolitical goals and illustrated perfectly the lofty presumptions at the heart of US Cold War nationality. America's global leadership rested on its purportedly peerless character and capabilities, which were evident in these technology intensive programs of scientific exploration. America's global leadership also rested on its special purpose. No purpose was as special, even transcendent, as the nation's declared effort to expand human knowledge and make the untapped bounty on earth and beyond serve the boundless aspirations of humankind. When national authorities promoted these programs by employing a salient nationalist myth, many keepers of public opinion followed their trusted lead by hailing the space and Antarctic frontiers in print, on screen, and over the airwaves. Just like the popular Disneyland amusement park, which glorified the nation's pioneering roots in its Frontierland and saw the US mature among Tomorrowland's high-tech outposts on earth and in space, they cast America as a pioneering nation forged on its frontiers and poised to range across the planet and other worlds.[26] These were grand expectations indeed, and they fortified a nationalist paradigm that deemed America's unprecedented reach for global leadership critical to its welfare and consistent with its glorious history and providential purpose.

Diverging Paradigms of National Exceptionalism

The motif of the space and Antarctic frontiers became conventional because credible authorities used it to promote the costly US space and Antarctic programs. As chapter 1 indicates, they promoted spaceflight and Antarctic exploration because these programs dramatically represented national superiority in scientific research and technological innovation, both of which buoyed America's economy, military power, and international prestige during the Cold War. These authorities used the frontier motif to do so, chapter 2 explains, because it cast their favored programs as true to the nation's pioneering character and as proof that America's democracy and market economy unleashed people's innate talents, as no other system could, for the benefit of all humankind. Thus the frontier motif drew strength from and reinforced a Cold War nationalist paradigm that justified US global leadership. The United States assumed that leadership, according to this paradigm, because the freedom and prosperity Americans enjoyed had become global imperatives, the necessary antidotes to a worldwide totalitarian menace. The R&D complex the United States raised to contain that menace would be wielded in outer space and Antarctica in a patently peaceful manner. The visionary Americans who did so bravely carried on their nation's finest tradition as they set out to conquer those frontiers. These mostly white men would use America's preeminent science and technology to make nature do their bidding and support the liberty and material aspirations of all people.

The many Americans who spoke of the space and Antarctic frontiers were true nationalists who assumed this was a straightforward, true account of the nation. The United States, in their minds, was not a fleeting historical accident. It was a vital and lasting community of Americans who had realized age-old human aspirations and were endowed with a unique national character rooted in natural principles.[27] They believed Americans consecrated themselves to these principles, shared the same hallowed history and geography, and faced a common and promising future. American Cold War nationalists differed from their more chauvinist predecessors in that they boldly proclaimed their country's commitment to freedom-loving people everywhere, not simply in the United States, for they all shared that liberal character and were entitled to that hopeful future. Outer space and Antarctica provided the perfect geography for this *internationalist* nationalism, for these vacant realms belonged to no one and thus could be made to serve everyone.[28] They were supranational regions without indigenous populations, places the United States demonstrated its innate knack for scientific research and technological innovation in a manner that purportedly benefited all humankind.

Proponents may have used the frontier motif to exalt their nation's timeless virtues and justify its global leadership. But the motif itself was not timeless, and it did not describe an objectively real and unchanging nation. It reinforced a historical paradigm that imagined the United States in a timely and culturally resonant way. The frontier motif helped curry public favor for the US space and Antarctic programs, and it suited the prerogatives and international ambitions of the Cold War nation-state and its many supporting constituents. The character and lifecycle of this *nationalist paradigm* were similar to those of the *scientific paradigms* famously described by historian of science Thomas Kuhn.[29] Just as nationalists tend to regard their nations as elemental and fundamentally unchanging, scientists often treat natural phenomena as concrete and knowable. As members of an intellectual community, Kuhn explained, scientists internalize as true particular ways of understanding those phenomena. These ways become that community's conceptual norms, and they sustain the prerogatives of scientists whose stature and research are oriented around those paradigms. Culturally and professionally invested in them, scientists tend to cling to these common held theories even if mounting data suggest the need for different, more accurate ways of understanding natural phenomena. Scientists often resist such change, Kuhn wrote, and try to adjust their prevailing paradigm just enough to accommodate contradictory evidence. If that paradigm is overwhelmed by such evidence, the scientific community might finally discard it for a new paradigm, a whole new world view that better accounts for that data.

Such was the case with the nationalist paradigm that turned outer space and Antarctica into frontiers for the American Century. The profound changes the United States went through in the late 1960s essentially overwhelmed that paradigm and its underlying worldview. Those changes weakened its cultural salience, making its racial and gendered tones anachronistic. If the once benevolent picture of white male explorers came off as racist paternalism, their high-tech conquest of distant frontiers seemed outdated as Americans grew tired of martial pursuits during the Vietnam War, expressed deep concern for the natural environment, and questioned the power of science and technology to positively transform the world. American culture changed dramatically, as did the prerogatives of the nation-state, which no longer enjoyed global economic preponderance. The United States could no longer afford a costly competition with the Soviet Union across the world and in outer space. It chose the more affordable policy of détente and accordingly tightened its presence in Antarctica and ratcheted down its expensive human spaceflight program. National leaders and program boosters found that the motif of the space and Antarctic frontiers had lost cultural footing and that it no longer conformed to a political economy oriented to Cold War accommodation and to economic competition with resurgent Western Europe and Japan.

As the frontier paradigm came under terrific pressure, the way that program boosters and public observers described the US space and Antarctic programs, and more generally the character and purpose of the Cold War nation, diverged over the next two decades. Antarctic exploration had never captured people's imagination quite like space travel, so the relatively small community of American scientists and government officials most invested in the US Antarctic Program (USAP) ultimately discarded the motif of the Antarctic frontier. According to chapter 3, those scientists had been primarily concerned with basic research rather than frontier conquest, and arbiters of public opinion gave their voices greater prominence in the 1970s as Antarctic science promised important dividends. Some opinion makers thought that Antarctic researchers might discover swarms of marine life to feed a hungry world and oil and mineral reserves to sustain industrial civilization. Their hope that Antarctica would become a resource frontier competed with a vision increasingly preferred by south polar scientists and environmental activists. That vision raised Antarctica as an irreplaceable platform for scientific study of the global environment, a pristine wilderness where researchers could best assess the ecological crises facing humankind. As south polar geopolitics became oriented around environmental research in the 1980s, the prerogatives of US diplomats and defense officials similarly changed and they began describing the USAP as a leading effort to better understand the planet. Only thus could humankind avert global environmental catastrophes such as climate change and stratospheric ozone depletion. By the end of the Cold War, Antarctic specialists and public opinion makers agreed that the southern continent was a sublime and critically important wilderness, rather than a godforsaken frontier rich for the taking. They achieved what Thomas Kuhn called a paradigm shift. They still described the exceptional United States as the leading force in Antarctica for human good. But they did so in a way that was incommensurable, in Kuhn's words, with the frontier motif. They described a nation bent on environmental preservation rather than frontier conquest, a people committed to repairing instead of improving the natural world.

Conquering the space frontier became similarly dated, chapter 4 explains, when federal officials facing economic crisis dramatically trimmed NASA's budget. As national excitement about an American Century gave way in the 1970s to the realpolitik of détente and to global economic competition, those officials dropped the lofty frontier motif and reoriented a shrunken space program to providing practical benefits to people on earth. Their more instrumental plan for America's space age future entailed technology development and satellites for global communications, weather analysis, resource surveys, and environmental assessment. This prosaic agenda was certainly important and suited the nation's downsized expectations and finances, but it did not pack the visionary luster and celestial cant of the space frontier. As powerful aerospace interest groups adapted to new political and economic constraints, conditions were ripe for a paradigm shift in which they promoted outer space, like Antarctica, as a workaday place where the United States demonstrated its deserving world leadership by helping humankind live within earth's constraining limits.

That shift did not occur. There were too many interests committed to a more robust spaceflight program as well as Americans who remained deeply attached to the transcendent idea of the space frontier. They kept the dream of the space frontier alive during the 1970s and helped revive the paradigm of spacefaring nationality in the next decade, when the Cold War heated up again and many national leaders promoted a visionary spaceflight program as a necessary means to recover America's global stature. With a fleet of space shuttles in operation and the much anticipated "Space Station Freedom" on the drawing board, a presidential commission expressed the starry-eyed optimism of many Americans during the 1980s when it called on the United States to use these assets to chart a lasting path into space. As its report *Pioneering the Space Frontier* plainly indicated, many people once again regarded outer space as a frontier for the American Century.[30]

Conservative pols and spaceflight enthusiasts likely welcomed this as a fortuitous return in which Americans remembered their nation's true nature and grand destiny after an aberrant detour in the 1970s. But the concluding chapter argues that they would have been mistaken, for the nationalist paradigm of a spacefaring America faltered once again at the end of the Cold War. That paradigm had invoked the nation's righteous leadership in a bipolar world. It did not offer a credible profile of America's position in a multipolar world defined by earthly competition among capitalist economies, including those of ex-communist Russia and China, and by regional threats from so-called rogue nations. Nor was that nationalist paradigm sustained by attacks from radical groups such as al Qaeda and more recently the so-called Islamic State (IS). The United States has been embroiled in a "War on Terror" against such groups, but their rejection of secular progress has not incited a return to the space frontier. The Soviet promise of a glorious life on earth and in space made possible by communism warranted an American response and aroused spacefaring nationalism in the United States. Americans have not responded in that way to the theocratic vision among suicide bombers and IS insurgents of the heavens as paradise for fallen martyrs.

The utopian dream of the space frontier—American democracy and capitalism literally delivering new worlds—also faltered owing to the country's lack of money and Americans' eroded faith in the transformative power of science and technology. A nation in chronic budget deficit could ill afford that dream, and many Americans questioned whether their limited technology, which spectacularly failed in space shuttle disasters in 1986 and 2003, could sustain them in the forbidding realm of outer space. The motif of the space frontier suffered the further blow of ideological change in the United States. Neoliberal proponents of privatization and individualism undercut an essentially liberal paradigm that championed government coordinated action on earth and in space for the greater good of all people. That paradigm has given way to gathering hostility to government and taxation, a critical source of funds for NASA and the US spaceflight program. That liberal paradigm has also given way to cost-cutting outsourcing of many federal services, including some provided by the space agency. For instance, private firms are now building rockets and spaceships that are not only replacing NASA's recently retired shuttle fleet but will also open the door for more multimillionaires to launch into space, visit the International Space Station (ISS), and even vacation in orbit. Once marketed as a steppingstone to the space frontier that would generate new knowledge and technologies of universal benefit, the ISS has come to symbolize what the *Economist* magazine calls "The End of the Space Age." It represents the orbital limits of America's space ambitions and the commercialization of international spaceflight, most visibly on behalf of those wealthy tourists.[31]

A market-driven spaceflight program does not evince the lofty purpose and effective national action of the American Century. The same is true of a US Antarctic Program oriented to environmental research and protection rather than the polar conquest on display at the 1964 New York World's Fair. The aims and public representations of the US space and Antarctic programs underwent significant revision in the decades after that World's Fair because the cultural politics of federal undertakings dramatically changed. The *political culture* of Washington endured, in that interest groups and program stakeholders continued to encourage support from taxpayers and elected representatives for their favored federal projects. NASA and NSF officials once called this enhancing "public understanding" of the US space and Antarctic programs.[32] However the *cultural politics* of these programs, the culturally meaningful terms these groups employed to garner that political support and influence public understanding, changed in the late twentieth century in response to dramatic shifts in

American domestic and international affairs. The sudden emergence and long decline of the motif of the space and Antarctic frontiers paralleled those shifts and reveals that a once strong Cold War consensus, and particularly people's hegemonic belief in the American Century, broke down in the late 1960s. This breakdown was not a sign that nationalism had become "historically less important," in the late historian Eric Hobsbawm's words. [33] Instead nationalist conceptualizations of American character and position in the world became more varied and contested. Those multiple conceptualizations became more contested as Americans worried that the nation had lost its footing and wondered how the United States could recover its power, prestige, and purpose. The country's commentariat thereafter saw in the US space and Antarctic programs two very different profiles of America. Together with program advocates, they ultimately exalted the nation as an environmental steward in Antarctica but continued to tussle over the framework and purpose of US spaceflight, which many still regard as the best hope for humankind, its path to a starry frontier.

In accounting for the sudden rise and long decline of these *Frontiers for the American Century*, this study offers a detailed overview of late-twentieth-century US history. It does so with an analytical emphasis on American culture, specifically on prominent arbiters of national opinion who weighed in on US policy in Antarctica and outer space. That emphasis does not fall in the postmodern camp which argues that public culture has no fixed center since all individuals perceive their world in unique ways. Granted, each person's viewpoint is unique and an ongoing proliferation of media outlets has splintered public discourse. But Americans framed their perceptions about national power and purpose during this period with information and editorial opinions drawn from a shared public sphere. This study focuses on that sphere and on the national authorities who earnestly fed it information and opinion. In addition to promoting their narrow interests and US activities in outer space and Antarctica, their discourse influenced people's impressions of the world and America's preferred place in it. This influence was likely particularly strong in these instances, for most people learned about the US space and Antarctic programs from respected public authorities rather than through direct personal experience. Those authorities initially engaged in a cultural project essential to America's Cold War power and stature. They invoked hegemonic conceptions of national character and purpose to promote those strategic programs and cast them in a favorable light. Their voices illustrate the abiding importance of these programs, and more generally of America's civilian and military R&D complex, to Cold War national power and culture.

By comparing the cultural politics of the US space and Antarctic programs, this study encourages students of modern America to see how these programs, and science and technology more generally, were critical pillars of Cold War national identity. It shows how that identity was an essential means of building support for ambitious public policy during this period of unprecedented national power. This book also provides scholars of science and technology studies with an example of the reciprocal influences between national culture and federal R&D. Such historical analysis of the cultural politics of science and technology is wholly absent from the relatively few accounts of United States activity in Antarctica.[34] It is more evident in the enormous body of work on US space exploration.[35] However that work rarely analyzes the complex and changing character of the frontier motif, which many scholars recognize was a common framework proponents used to describe and justify the US spaceflight program. Lastly, none of those histories use the comparative case of Antarctica to reveal that paradigm of frontier nationality, which once framed public thinking about spaceflight and still lingers in the public imagination, can shift if national prerogatives and culture decisively change. Nor do they use the case of outer space to illustrate the strength and inertia of nationalist identity. The frontier motif still appeals to many Americans who see visionary national action in outer space as key to insuring that the twenty-first century remains an American Century.

Chapter 1

Rising to the Sputnik Challenge

On the evening of October 4, 1957, the Soviet embassy feted specialists gathered in Washington, DC to coordinate rocket and satellite launches for the unfolding IGY. Well briefed about America's IGY satellite plans and still uncertain about the progress of Russia's program, most guests expected the United States to reach earth orbit first in as little as two months. They were understandably surprised then when Lloyd V. Berkner, the American member of the international IGY coordinating committee, begged everyone's attention to pass on some stunning news: their hosts had just orbited Sputnik, the world's first artificial satellite. The assembled scientists and engineers raised their glasses and applauded this groundbreaking achievement.[1] Their ovation was not simply the good etiquette of embassy guests. They cheered Sputnik as a milestone for their professions and a historic juncture for humankind. Closing the door on millennia of earth-bound star gazing, Sputnik heralded a new epoch of cosmic exploration and even human settlement of other worlds.

Americans were generally not so high-minded. Unable to put aside national loyalties, many saw the beeping orbiter as a potent threat to United States security and standing in the world. Their ensuing public debate ranged far beyond the vague menace of Sputnik itself and the strategic threat posed by enemy rockets. They regarded the satellite as a worrying sign that the communist Soviet Union was not intrinsically backward and that it could, in fact, coercively marshal its resources and effectively compete with America. These Americans anxiously regarded the satellite as a Soviet accomplishment that had positively impressed the world and reflected poorly on the United States. As a prominent rocket designer later wrote, after Sputnik "it became popular to question the bulwarks of our society; our public educational system, our industrial strength, international policy, defense strategy and forces, the capability of our science and technology. Even the moral fiber of our people came under searching examination."[2]

Lloyd Berkner may have been one of the high-minded professionals who toasted his Russian colleagues, but he was also an influential science advisor who wished to defend the bulwarks of his society. He had long argued that America's leading research and development (R&D) was good for the country's scientific and engineering enterprises and paid great dividends in the form of enhanced national security, economic vitality, and international prestige.[3] Sputnik exposed the potential bitter fruits of this widely accepted formula by demonstrating that spectacular feats of science and engineering could improve the international standing of the Soviet Union as well. Worried about the global fallout from Sputnik, Berkner admonished politicians to redouble federal support for R&D and create "clear symbols of intellectual leadership" to inspire Americans and impress the world. He identified scientific exploration as such a symbol and asserted that in an era "when few physical frontiers remain, peoples visualize space and the Antarctic…as challenges that must be accepted by a great nation to demonstrate its mettle." The United States must have the largest footprints in these regions, he concluded, because "people expect this of us as the leader."[4]

It is not surprising that Lloyd Berkner came to this conclusion. A veteran Antarctic researcher and national authority on space policy, he was also a chief architect of the 1957–58 IGY. He knew Americans had designed their ambitious IGY projects in Antarctica and earth orbit not simply to advance geophysical sciences but also to bolster their nation's defense, economy, and reputation. Berkner looked to these existing programs to fulfill their original purpose and help the United States right its faltering prestige. After the IGY, federal officials did so as well. They charged the National Science Foundation (NSF) with managing the now permanent US Antarctic Program and they created the National Aeronautics and Space Administration (NASA) to direct America's ongoing civilian space program. The imprint of Sputnik remained strong during the ensuing decade as NASA and the NSF managed these leading exploratory programs so that they paid practical dividends, sparked domestic pride, and enhanced US prestige.

Science and Technology in Postwar America

In 1962 the Nobel Prize winning chemist and Chairman of the US Atomic Energy Commission Glenn T. Seaborg hailed science as America's "Third Revolution." If its eighteenth-century revolution for independence endowed citizens with liberty and democracy and its nineteenth-century industrial revolution "expanded the means for a happier life visualized by the Founders of the nation," then twentieth-century American science constituted a third revolution that further "raised the stature, dignity, and capacity of humanity." He praised American scientists for helping cure debilitating diseases, improve agricultural yields, generate new and more productive industries, and defend the nation with high-tech weapons. They were able to do this, Seaborg explained, because science had "moved from the periphery to the center of society" since World War II.[5] It was no longer the preserve of detached academics and a few industrial laboratories. Science had become a strategic resource intensively cultivated by the federal government. Universities and industrial corporations remained important patrons, but now that national leaders deemed cutting-edge research essential for military security and economic growth, international aid and trade, and standing in the world, they no longer assumed a laissez-faire posture. Washington invested heavily across the country in university education, industrial capacity, and in-house government expertise and made path-breaking science and technology foundations of the "American Century."

When publisher Henry Luce coined the term in early 1941, he held out the dream of an American Century as reason for the United States to enter the war then engulfing the world. If it remained on the sidelines, he warned, fascist empires would dominate Europe and Asia. They would isolate the country, brutally subjugating the overseas friends and trading partners America needed to brace its liberal democracy and free enterprise system. Luce avoided blaring the trumpets of war, but he implored his fellow citizens "to accept wholeheartedly our duty and our opportunity as the most powerful and vital nation in the world" and act as the world leader. Only through their unfaltering world leadership could Americans bring about a novel era in which a "vital international economy and...moral order" sustained "the freedom, growth, and increasing satisfaction of all individual men." Harbingers of Luce's "revolutionary epoch" appeared within months as President Franklin D. Roosevelt signed the Atlantic Charter, which called for a peaceful postwar order of sovereign trading nations, and then the country went to war against the Axis Powers. At the end of hostilities, the rough outlines of an American Century fell into place as the United States became the primary force behind a United Nations for international diplomacy, economic institutions to facilitate free global trade, and multilateral organizations to achieve regional security. As Luce predicted, America's "revolutionary" advances "in science and in industry" gave the United States the confidence, power, and prestige they needed to initiate this new epoch.[6]

America's world leadership depended foremost on a robust economy, one recently stoked by scientific research and technological development. Just a few years earlier many economists dourly expected the United States would never fully recover from the Great Depression. They believed America's free enterprise system had reached its natural limits and would no longer enjoy what they deemed were the factors that had previously driven economic growth: new geographic frontiers; population expansion; and industrial innovation. Now that its "mature economy" would no longer expand via free market forces alone, the federal government had to help balance national production and demand by investing in public works, facilitating collective industrial agreements, and sometimes setting wage and price controls.[7] These plans for federal management of an exhausted economy fell out of favor when the United States boomed once again during and after the war. Science and technological innovation had less to do with this economic resurgence than government spending, swelling domestic consumer demand, and booming exports to support overseas reconstruction. Nevertheless many experts believed that ongoing federal support for R&D would complement America's free enterprise system and help Washington fulfill its commitment, outlined in the 1946 Employment Act, "to promote maximum employment, production, and purchasing power."[8]

That R&D did not create new geographic frontiers. But the geologists and engineers who charted and dammed America's western rivers in the 1930s and who surveyed and mined valuable new deposits there during the war demonstrated that modern science and technology could make old frontiers turn a new profit. After years of declining birth rates, the postwar "Baby Boom" triggered the population growth and swelling consumer demand experts had long deemed critical to a vital economy. Once again researchers and technicians lent a hand, designing more efficient systems of production that contributed to the rising wages and affordable prices that allowed a burgeoning population to snap up goods and services. Dramatic leaps in agricultural and manufacturing productivity stimulated the economy, which was also invigorated by new products and industries. R&D unplugged previously sluggish markets after the war. Chemical corporations sold novel pesticides and plastics, pharmaceutical companies marketed breakthrough drugs and vaccines, and aircraft manufacturers delivered planes that flew faster and farther than any before. With brand new research-intensive industries dedicated to high-tech computers, rockets, and atomic energy also emerging, many people reasonably believed that greater federal support for science was needed to help sustain economic growth. Justifying this support, President Harry S. Truman highlighted the importance of "increasing in this country the vitality of the basic research upon which all technological development—and therefore our economic progress and national security—is dependent."[9]

The United States government had been an important sponsor of basic and practical research in the nineteenth century when federal agencies drafted maps that facilitated maritime trade, provided farmers useful tips based on new agronomical research, and surveyed the natural resources that fueled westward expansion and industrialization. Washington continued these activities in the industrial age, but several new institutions became the primary underwriters of American science at the time. Basic science took place largely in new research universities where faculty scientists and engineers trained growing numbers of graduate students by putting them to work as research assistants.[10] These teams often scraped by with university funding. Some enjoyed support from philanthropic organizations recently charged by rich industrialists with names like Rockefeller and Carnegie with elevating the tastes and talents of the American public. These organizations famously did so by building libraries, museums, and concert halls, but they also endowed private research institutes and financed some university laboratories. While university laboratories focused on basic research, industrial corporations became the primary benefactors of applied science. Many contracted college professors to conduct targeted research related to their industrial products and processes. Some of the largest enterprises invested heavily in their own labs, betting that in-house R&D would generate patent-protected products to give them legs up on their stiff competition. If industrial-age research developed largely within this loose federation of universities, philanthropies, and industrial corporations, the center of gravity of the scientific enterprise swung back toward the federal government during World War II.[11]

Of course the overriding goal of US support for science at the time was to produce machinery and armaments to win the war. Built from existing materials and technology, the bulk of that stock flowed like a torrent from corporate drafting rooms and industrial assembly lines. Some of the more sophisticated materiel emerged instead from research laboratories. The federal Office of Scientific Research and Development (OSRD) channeled several hundred million dollars of wartime government R&D funds to these labs, particularly those working on cutting-edge electronics. It supported Massachusetts Institute of Technology's Radiation Laboratory, for example, which designed many of the radars that tracked Axis bombers and made the Atlantic Ocean treacherous for German submarines late in the war. OSRD money went to the Johns Hopkins University's Applied Physics Laboratory, which invented the proximity fuses that detonated antiaircraft shells with devastating effect near enemy planes. The agency also had a hand in the most significant R&D program to date. After the head of the OSRD and principal architect of wartime R&D Vannevar Bush recommended the plan, the United States started a crash program to build an atomic bomb before Germany did so. Managed by the US Army, the resulting top-secret Manhattan Project contracted corporations like chemical giant Du Pont and relied on a legion of technicians and scientists, including scores of the world's most prominent nuclear physicists and chemists, to erect what was the largest industrial infrastructure ever devoted to a single endeavor. The more than 100,000 people tasked to the nearly $2 billion Manhattan Project built heavily secured plants across the United States and rolled out the atomic bombs that obliterated large swaths of the Japanese cities of Hiroshima and Nagasaki.[12]

National leaders understood well that this enormous wartime R&D infrastructure also contributed mightily to the national economy. This was one of Vannevar Bush's key points in his seminal 1945 report *Science—The Endless Frontier*. Still at the helm of the OSRD, Bush urged the federal government to create a National Research Foundation (NRF) and maintain its high level of support for basic science. According to Bush:

> Advances in science when put to practical use mean more jobs, higher wages, shorter hours, more abundant crops, more leisure for recreation, for study, for learning how to live without the deadening drudgery which has been the burden of the common man for ages past. Advances in science will also bring higher standards of living, will lead to the prevention or cure of diseases, [and] will promote conservation of our limited national resources.[13]

Many critics of Bush's administrative outline for an NRF nevertheless agreed that federal support for science was essential to America's security and economic wellbeing. Democratic representative Harvey Kilgore from West Virginia, for instance, rejected Bush's pitch that an NRF should fund only the most qualified researchers in the basic physical sciences. Kilgore felt the United States would better profit from an NRF that distributed money more equitably to researchers across the country as well as to social scientists working on pressing national problems. Truman advisor John Steelman and fellow authors of the 1947 report *Science and Public Policy: A Program for the Nation* also valued the practical benefits of basic science. They criticized Bush's proposed NRF as insulated from White House oversight and felt a president-appointed director would keep such an institution politically accountable so that it would intentionally serve the public good.[14]

The resulting NSF was an administrative hybrid of the Bush and Steelman reports; its director was appointed by the US President, while panels of prominent specialists awarded NSF grants to the most qualified applicants conducting largely university-based basic research in the physical sciences. The new agency embodied the common assumption that pure science brought forth new knowledge that applied scientists and engineers needed to create revolutionary new technologies. But the NSF was not the sole or even primary source of federal research dollars. The National Institute of Health financed medical research, while defense-oriented agencies supplied most government funds for applied research and basic science. When the NSF opened its doors in 1950, the Office of Naval Research and the Atomic Energy Commission (AEC) were already the most generous federal patrons of scientific research. Together these agencies supported more traditional, so-called little science by helping individual faculty and university departments across the country equip their laboratories and expand graduate programs. Industrial enterprises, in turn, often employed their highly trained graduates and reaped lucrative government contracts to develop weapons based on university research. The AEC in particular also supported the relatively new phenomenon of "big science." Coming of age in wartime weapons R&D programs, most notably with the Manhattan Project, big science was very costly and operated at an unprecedented scale, employed expert staff from many academic disciplines, and often required large and high-tech apparatus, laboratories, or field stations.[15] Bankrolled by the federal government, big science projects produced data that engineers used to design powerful weapons and groundbreaking civilian technologies. Favorably comparing these grand projects to engineering marvels of past civilizations, the AEC's Alvin Weinberg neatly summarized a convention among big science practitioners when he asserted: "Society could hardly survive for many more generations without the fantastic developments that have come out of big science."[16]

Although AEC big science projects harnessed nuclear energy primarily for weapons development, Congress mandated that it also uses that energy "toward improving the public welfare, increasing the standard of living, [and] strengthening free competition in private enterprise."[17] The AEC attempted to do so by bankrolling its academic and industrial partners to study and develop commercial applications of atomic energy. These efforts were part of a swelling "military-industrial-academic complex" that supported little and big science and exerted a powerful influence on America's postwar economy.[18] Some critics worried that complex diverted too much money and talent away from civilian enterprise and into military research and weapons production. Nevertheless it was a force for economic modernization by strengthening American universities, deepening the national pool of scientists and engineers, and providing the rich government contracts that industrial firms used to update factories and diversify operations.

If science policymakers deemed path-breaking American R&D as critical to domestic economic health, they also saw it as an important tool for modernizing the global economy. To secure international peace and prosperity through American global leadership, White House and State Department officials pursued what historian Melvyn Leffler labeled the "grand strategy" of building "a world trading system hospitable to the unrestricted movement of goods and capital."[19] They believed this trading system would prevent the insular economic nationalism and imperial conquest that had triggered two world wars, and it would contain the socialist policies pushed by the Soviet Union and embraced by many Europeans, Japanese, and postcolonial nationalists. It would do so by tying independent nations together through growing and universally beneficial trade and cross-border investment. Each nation would see its fortunes rise if it avoided autarky and socialism and instead opened its free market to the world in a way best suited to its level of development. The least developed countries would focus on exporting their natural resources. Semi-developed nations would process their natural resources and assemble basic industrial goods. The United States and its industrial partners would hold the system together by supplying critical investment to less developed countries and producing advanced equipment, consumer goods, and professional services for domestic and foreign consumption. Each nation would purportedly benefit through economic specialization and interdependence, and Americans stood ready to use their superior science and technology to help them achieve these designated specialties and become vested in a free and prosperous global civilization.

According to this ambitious scheme, the poorest nations desperately required American expertise and equipment to properly exploit their natural resources. As President Truman declared in his 1949 inaugural address, the United States "must embark on a bold new program for making the benefits of our scientific advances and industrial progress available for the improvement and growth of underdeveloped areas."[20] Whether that new program was delivered as aid or paid for through trade, it entailed American engineers and heavy machinery helping underdeveloped nations build modern infrastructures for efficient natural resource production. Scientists would also play a role by teaching people in these countries to transition from traditional to more profitable forms of agriculture, forestry, and mining so that they could finance further development and improve their standard of living. Truman had little to say about semi-developed countries, but they too required American assistance to effectively develop their natural resources. Because these nations also fit into the international economy as basic manufacturers, they relied on American know-how and equipment to train local technicians and build the factories that processed raw materials and produced lower profit goods spun off from advanced industrial countries.[21] The president did mention these latter countries in his speech, particularly those in Europe, and promised American economic cooperation so that "the free people of that continent can resume their rightful place in the forefront of civilization and can contribute once more to the security and welfare of the world."[22]

Although Truman then offered few details about this cooperation, his administration counted on using US science and technology to fortify liberal allies in Europe.[23] The United States provided high-tech farm machinery, factory apparatus, and utility equipment to war-torn Western European countries. Dislodged from many of their former colonies and separated by an Iron Curtain from their traditional Eastern European trading partners, these countries needed such assistance for the reconstruction necessary to resume their places at "the forefront of civilization" as industrialized capitalist nations. American researchers helped by building strong working relationships with Old World colleagues. US colleges trained many of their young scientists and engineers.[24] Corporate enterprise contributed too by building manufacturing plants in Western Europe and licensing advanced American technology to European companies. Proponents of America's "grand strategy" believed this aid, trade, and professional collaboration would turn a revitalized Western Europe into a pillar of the so-called Free World and a liberal bulwark against the Soviet Union.

During the first decade of the Cold War, American statesmen were preoccupied with checking Soviet influence in that strategic corner of the world. In addition to scientific collaboration, commercial technology exchange, and billions of dollars of Marshall Plan economic assistance, US foreign policy experts saw America's matchless military R&D as critical for maintaining the cross-Atlantic alliance. The willingness of wartime belligerents to attack civilian population centers combined with the terrifying atomic weapons that mushroomed out of World War II left people on both sides of the ocean convinced that any future war would lay waste to Europe. During its four-year atomic monopoly, the United States endeavored to prevent that destruction by holding these weapons and its technically superior air and naval forces as a counterweight in Europe to the Soviet Union's formidable armies. The challenge of keeping Americans and their European friends dedicated to that alliance deepened at the cusp of the 1950s after the USSR became an atomic power, communists consolidated control of China, and Moscow's ally North Korea invaded South Korea. World-weary Americans now had their worst fears confirmed; emboldened with atomic arms, their Soviet adversary seemed to be militantly building its influence across a volatile world. Western Europeans still weakened from war had to take these events into account and consider whether their national interests, even their survival, required accommodating the communist superpower on their eastern borders.

The Truman Administration had already deflected homegrown resistance to postwar internationalism by whipping Americans' moral disgust with communism into a galvanizing political force, one it used to curry domestic support for US leadership of the "Free World."[25] Now their gathering fear of Soviet power abroad and ideological subversion at home, fueled by sensible anticommunists as well as political demagogues like Senator Joseph McCarthy (R, WI), left many alarmed Americans ready to protect themselves and their Free World allies through greater military spending. As US armed forces budgets effectively doubled between 1950 and 1952 and exceeded a muscular ten percent of gross domestic product for much of the decade, policymakers channeled billions into military R&D programs.[26] With the success of wartime science fresh in their minds, security experts counted on vigorous military R&D to spawn the superior weapons needed to deter a Soviet adversary with larger armies and technically advancing armaments. In a sustained program that dwarfed the Manhattan Project, which built atomic bombs capable of smashing urban centers, the United States developed and stockpiled the far more powerful hydrogen bombs that could annihilate whole cities. It also produced long-distance bombers, shorter-range rockets, and atomic-powered submarines to deliver these thermonuclear weapons, and erected an early warning radar system across Alaska and northern Canada to detect nuclear attacks from the Soviet Union.

America's costly high-tech military edge failed to rollback communist gains in Eastern Europe and Asia. But it helped the United States achieve its grand strategy of keeping Western European countries in its camp. Leaders there were justifiably wary of the USSR, which used its formidable armies in the late 1940s to suppress western-leaning governments in Eastern Europe and blockade the divided German city of Berlin. US resistance to that blockade and its subsequent pledge as a member of the North American Treaty Organization to defend its Atlantic partners allayed European worries, and its swelling high-tech arsenals helped deter Soviet incursions further west. This emboldened many Western European leaders to ally with the United States. Washington's promise to entrench its military forces throughout the region also gave liberal reformers there the confidence that a reindustrialized West Germany, pushed by the United States as necessary for the recovery of Western European capitalism, would not threaten its neighbors once again.

Thus national officials wielded what scholar Joseph Nye, Jr. called the "hard power" of America's "economic and military might" around the world, and particularly in Western Europe, during the Cold War. Advanced science and technology bolstered that hard power, but they also facilitated US foreign policy by reinforcing what Nye dubbed "soft power." To entice other nations to willingly adopt its strategic goals, the United States built up the soft power of international respect that came from its credible strength and attractive principles.[27] Using more common terminology decades earlier, Henry Luce identified this "prestige" as critical to the American Century. "American prestige throughout the world," he asserted, "is faith in the good intentions as well as in the ultimate intelligence and ultimate strength of the whole American people."[28] Their unrivaled R&D positively reflected the intelligence and strength of the American people. As vehicles for military defense and economic modernization, science and technology featured prominently among government officials' declarations of America's designs for a secure and prosperous Free World. Those officials also cultivated scientific cooperation with America's allies so as to strengthen its global relationships and cast a favorable light on its internationalist agenda.[29] As one historian of science policy noted, "national prestige" was "often the motive behind cooperation" since Americans' exchange of scientific data with professional colleagues abroad "was widely admired."[30] Most important, public officials reckoned that because good science and engineering occurred only when practitioners enjoyed the liberty to develop their individual insights and vet their findings democratically with one another, America's political ideology and free and open society accounted for its leading science and technology. These officials hoped America's superior R&D would bolster its prestige by showing clearly that life-improving science and technology flourished in the

United States and spread rapidly to its allies while they simply hobbled along in the stultifying, authoritarian Soviet Union.

As Glenn Seaborg noted, postwar science had become a strategic resource that served American interests well. Statesmen, science policymakers, and national security advisors took these interests into account when they backed strong US participation in the IGY of 1957–58. Unprecedented in its scientific scope and scale of participation, the IGY was devoted to basic geophysical research and became a model of peaceful scientific internationalism. US officials, like those in many participating countries, applied a more mercenary calculus when planning its IGY program. With America's hard and soft power in mind, they decided that its security, economy, and prestige, as well as basic science, would best be served if the United States mounted a leading IGY program, especially in the forbidding realms of Antarctica and outer space.

The International Geophysical Year

At a dinner party of distinguished geophysicists in April 1950, Lloyd Berkner planted the seed of the IGY. Fellow party-goers agreed with him that the time had come to update their profession's paltry store of data on the planetary environment. Their initial proposal to the International Council of Scientific Unions (ICSU) was to model the First Polar Year (1882–83) and Second Polar Year (1932–33) and undertake a multinational research program around the North and South Poles. When member scientists lobbied ICSU to set its sights beyond a Third Polar Year, that scientific body formed a special committee to coordinate the now IGY's worldwide observations of the earth and its oceans, atmosphere, and near-space environment. Berkner and his colleagues on the coordinating panel, the Comité Spécial de l'Année Géophysique Internationale (CSAGI), felt the global networks of scientists, sufficiently built up since World War II and equipped with new and powerful observational tools, were well prepared to gather that geophysical data. They relied on international scientific congresses and their national affiliates to design and implement IGY research projects. The program's prime movers counted on the governments of a whopping 66 countries to support their contingents of the more than 10,000 participants by paying for expensive apparatus and transporting them and their equipment to remote locations. CSAGI settled on the 18-month stretch between July 1, 1957 and December 31, 1958 for the IGY since this spanned an anticipated period of maximum solar activity. Whereas Second Polar Year scientists observed the relatively quiet part of the sun's energy cycle known as the solar minimum, IGY researchers planned to study how heightened solar storms affected earth's physical environment.[31] The IGY was a turning point in modern science. It built a huge stock of geophysical information and established institutional partnerships for international scientific collaboration that continue to this day. The people who organized and participated in this tour de force were plainly motivated by their love of science and desire to work with foreign professionals. A prominent American scientist evoked this idealism by describing the IGY as an opportunity for "men of all nations tired of war and dissension" to turn "to mother earth for a common effort on which all find it easy to agree."[32] Another went further and called the program "the single most significant peaceful activity of mankind since the Renaissance and the Copernican Revolution."[33] President Dwight D. Eisenhower was only slightly less gushing when he called the IGY "one of the great scientific adventures of our time" and a "demonstration of the ability of peoples of all nations to work together harmoniously for the common good."[34] Eisenhower stuck this idealistic note, but he also knew very well that these history-making people were able to turn to mother earth when keepers of the public purse deemed it prudent to do so. After all, scientists relied on government officials who paid close attention to more mundane factors of national interests to pony up the IGY's collective $2 billion price tag.[35]

Such political calculus was not plainly visible since the national interests of participating states were well served by the IGY's program of basic research. Still government representatives likely recognized the bankable prestige that came from sponsoring IGY projects. Their superpower counterparts certainly did. Rightly suspicious about a paranoid leadership that had cloistered Soviet scientists, the *New York Times* derisively quipped that "one of the chief reasons for full Soviet participation in the International Geophysical Year was to increase Soviet prestige in science."[36] The newspaper's editors could have looked closer to home, for US officials also had national prestige in mind as they planned for the IGY. Again, these officials saw scientific leadership as America's bailiwick which established its credibility as the military defender and economic cornerstone of the Free World. This leadership confirmed Americans' commonly held belief that the scientific enterprise thrived best in liberal democratic societies. US officials also felt that its outstanding efforts to stimulate scientific internationalism established productive relationships with other countries and cast a favorable light on its broader agenda of political and economic internationalism.[37] When lobbying for money in Washington, American scientists exploited these points and let on that the nation's standing would suffer if its IGY program was inferior to that of the Soviet Union.[38] White House and congressional authorities concurred and released the money for America's leading IGY program.

The United States spent hundreds of millions of dollars because these authorities also expected that IGY research would have great economic and military value. IGY meteorologists intended to collect data on high-atmosphere conditions and jet-stream winds needed by America's burgeoning commercial airline industry. Atmospheric scientists planned to examine ionospheric fluctuations triggered by solar storms that periodically disrupted electrical transmissions and radio and television broadcasts. Since mineral and petroleum companies used magnetic anomalies to find new resources, the *New York Times* then reported, they stood to profit as well from IGY scientists' global geomagnetic surveys.[39] National security advisors also heartily approved of US support for the IGY. As a Navy Admiral noted, the IGY's "coordinated world-wide effort in the main fields of atmospheric physics can be expected to yield basic information not only of general technical value but of value to our national defense."[40] A classified report further determined that without the IGY's "synoptic observation of certain geophysical phenomena of increasing relevance to the Department of Defense," such as atmospheric properties that affect ballistic missiles and oceanic attributes that influence submarine warfare, the US armed forces would otherwise have to make "such observations in its own behalf and at its own expense."[41] Despite publicists' emphasis on "the independent and purely scientific character of the American IGY programme," one historian nicely sums up, it was geared "to United States national interests and to relevant parts of the administration's national security policy."[42]

National leaders decided to send a huge expedition to Antarctica for the same reasons they supported the IGY as a whole, to encourage basic scientific research and enhance America's economy, prestige, and security. They designed America's costly expedition to serve ICSU's goal of collecting data across Antarctica. The scientific importance CSAGI afforded the area was evident in its programmatic criteria. In addition to "concurrent synoptic observations at many points," presumably including the southern continent, CSAGI endorsed "observations of all major geophysical phenomena in relatively inaccessible regions of the earth" as well as "epochal observations of slowly varying terrestrial phenomena."[43] The distant ice-choked continent certainly fulfilled the former requirement, while the latter favored studies of Antarctic ice sheets, whose size, and future growth or recession would affect global climate and sea levels. The organizing committee's desire that participants conduct simultaneous measurements at all latitudes also meant the southern continent had an important role to play in the IGY. Antarctic bases were needed to complete the chains of hundreds of observation stations strung from Pole to Pole along three longitudinal meridians.

Now that naval ships, helicopters, heavy-lift aircraft, and radio communications had at long last opened the region to comprehensive exploration, CSAGI pointedly determined "that Antarctica represents a most significant portion of the earth for intensive study during the International Geophysical Year."[44] As one of a dozen countries that answered CSAGI's call, the United States set up an impressive constellation of stations around the largely uncharted continent that made significant contributions to many fields of IGY research. American seismologists and their international colleagues assembled a general picture of the south polar ice sheets, which were far larger than they predicted, and discovered that under the ice Antarctica was not a contiguous landmass. They determined that East Antarctica was a solid landmass, West Antarctica was an archipelago, and portions of the continental crust in both regions were so depressed by vast ice sheets that they laid below sea level. Glaciologists set an important precedent for future research on global climate history by using a 1,000-foot ice core, taken at America's Byrd station, to estimate the area's average temperature and precipitation over nearly 2,000 years. Meteorologists from the 12 expeditionary nations conducted the very first continental surveys of Antarctic weather. They filed their data at "Weather Central" located in the Little America Station, which then radioed those findings to climate modelers around the world. At the geographic South Pole Station, Americans worked under extremely challenging conditions to map the magnetic field overhead, record electromagnetic pulses that bounced from Pole to Pole through space, observe cosmic rays streaming down through a gap in the soon-discovered Van Allen Radiation Belts, and photograph southern auroras as they brilliantly flared during solar storms.

Its prime movers publicly identified scientific exploration as the sole function of America's Antarctic expedition. They surely had economics in mind as well since public officials and national media had long expected Americans to turn a tidy profit in Antarctica. They had done so for well over a century as America sealers and then whalers returned from the area heavy with precious takings. When Washington dispatched two naval contingents to the region in the late 1940s, Metro Goldwyn Mayer's movie *The Secret Land* depicted them as forerunners of a new era of wealth acquisition in Antarctica. By describing these expeditions as a naval "epic of courage and sacrifice" whose ultimate goal was "the discovery and release to the world of the unknown treasures of the only untouched reservoir of raw materials left in the world," the movie borrowed a page from Rear Admiral Richard E. Byrd.[45] This veteran explorer had become a national celebrity for his pioneering flights over the North and South Poles and for his expeditionary leadership in what he dramatically described as the forbidding and desolate Antarctic continent. But Byrd saw that desolate icescape as rich in hidden treasures such as "coal and minerals" which he insisted "can ultimately provide support for the world economy." He also contended that the "preserving quality of the Antarctic cold" meant that surplus crops and perishable foods kept there would furnish "an international stockpile to help countries afflicted by famine or disaster."[46] Members of the US Joint Chiefs of Staff (JCS) quietly discounted Byrd's latter contention but advised the National Security Council (NSC) that researchers might indeed uncover valuable minerals in Antarctica and would certainly generate information critical to "shipping, aviation and radio communications of all sorts."[47] NSC and State Department specialists took this into account when they considered whether the United States should claim Antarctic territory. Although they remained cool to a territorial claim, one US Senator expressed the common feeling that "growing recognition of possible mineral values in the mountainous regions of that large continent make this important."[48]

IGY planners gave greater weight to national prestige and determined that an impressive Antarctic program would reinforce US credibility and confirm once again that its liberal democracy made scientific and technological leadership America's birthright. More important, they figured a substantial Antarctic expedition would secure prestige in the form of international deference the United States needed to achieve its strategic ambitions throughout the south polar region. By the eve of the IGY, America's primary goals there were to blunt territorial conflicts among its allies and limit Soviet influence in Antarctica. These foreign policy challenges did not exist when ships were still vulnerable to its menacing bergs and sea ice and when Antarctica remained terra nullius, a land without sovereign states. As modern naval technology made the ice encrusted continent more accessible, Great Britain filed the first claim over a portion of Antarctica in 1908, a wedge of territory ranging from an arc of coastline in pie-piece fashion to the singular point of the South Pole.

New Zealand, France, and Australia soon followed suit with similarly shaped territories centered on the South Pole, as did Norway, whose 1939 territorial claim came when German gunboats started prowling its traditional Antarctica whaling grounds. Seeds of political discord were fully set in 1940, when Chile and Argentina asserted priority over wedges of territory closest to the bottom tip of South America, including the stretch of Antarctic Peninsula already accounted for by Great Britain.

Richard Byrd regarded this state of affairs warily and urged military officials to substantiate United States rights in Antarctica by dispatching its naval power to the region. He argued that such a show of force, as occurred during US Operations Highjump (1946–47) and Windmill (1947–48), would also secure southern shipping lanes from Soviet aggression and prepare American troops for potential Cold War hostilities in the Arctic.[49] The JCS felt that in times of peace the southern continent "had little strategic value to the United States," but agreed that it might have such value in the event of global war with the Soviet Union. The JCS preferred that the United States save money and military resources by cooperating with its allies rather than flexing its naval muscles or asserting territorial rights in the region.[50] Washington followed this advice in 1948 and tried to improve souring relations among Great Britain, Argentina, and Chile and to keep the Soviet Union out the region by proposing that the United States and the seven "claimant nations" share sovereignty over the whole continent. This diplomatic push was resisted by several claimant states, which jealously defended their sovereign rights in Antarctica. It was also lambasted by the Soviet Union, which asserted the historic basis of its political rights in the region.[51] America's brief chance to fix the continent solely into its sphere of influence ended as Moscow blasted the "predatory tendencies of the imperialists of the USA in Antarctica," professed its "great economic and scientific interest in Antarctica," and demanded "complete observation of the principle of international cooperation in the decision of international problems" in that region.[52]

The IGY offered a temporary reprieve from this geopolitical knot. Looking to strengthen US influence and assure its access to the whole continent, the NSC determined in 1954 that scientific cooperation during the IGY was a way "to promote the over-all reduction of international friction, and the orderly solution of the territorial problems [in Antarctica] among friendly powers."[53] When the Soviet Union announced plans for an IGY Antarctic expedition, the United States and other world players stuck with this plan and judged scientific cooperation as the best means of postponing an uncertain political endgame in Antarctica. In their

1955 "Gentlemen's Agreement," 12 countries encouraged such cooperation and pledged not to put forward or abrogate territorial claims during the IGY.[54] The JCF secretly mulled over such a claim, but the United States worked from this agreement, in Secretary of State John Foster Dulles's words, for "some political status for Antarctica which will to the maximum extent possible exclude the Russians and assure that the area will develop under free world auspices."[55]

This objective framed America's Antarctic expedition. Although the United States sited its IGY stations there ostensibly for science, one student of international relations rightly called them political "wolves in the sheep's clothing of scientific research."[56] This was particularly true of the US South Pole station (Figure 1.1). Nearly a half century earlier, explorers Robert Falcon Scott, Ernest Shackleton, and Roald Amundsen raced to be the first to stand at the geographic South Pole, intent on personal glory and national prestige. US statesmen, defense officials,

Figure 1.1 Two men lower the US flag at America's South Pole Station, built for the IGY.

and IGY planners determined to bolster America's prestige by building a base at that remote and treacherous site. They also aimed to use that base to cement US political leadership in Antarctica. Since the station straddled all wedge-shaped territorial claims at their polar confluence, Washington could later use the base to justify priority over any part of Antarctica or to push for multinational cooperation there. Several prominent members of Congress recognized its value and discretely advised President Eisenhower to retain control of the base after the IGY.[57] Upon learning of Moscow's extensive Antarctic plans, the State Department also pushed to increase the number of American bases "to help offset any possible political consequences of Soviet action" on the continent.[58] US Navy Seabees built one of those stations to help Australia offset the strong presence of the Soviet Union, which built all of its IGY bases in that country's chosen sector. Although the relatively accessible Antarctic Peninsula was already well covered with IGY research stations, Seabees built yet another base nearby in case the United States needed diplomatic leverage in future disputes over that coveted territory.[59]

National prestige and security were similarly paramount in President Eisenhower's approval of a costly US program to orbit earth satellites during the IGY. Once again that effort was formally designed to advance basic geophysical research. The IGY took the whole planet as its object of study, and orbiting satellites promised the most comprehensive views to date of the global environment. Because satellites could observe what the NSF and National Academy of Sciences (NAS) called "extraterrestrial radiations and geophysical phenomena for extended periods of time" from outside earth's obfuscating atmosphere, CSAGI deemed they were of great scientific value and should be launched during the IGY.[60] When the United States took up CSAGI's call, White House Press Secretary James Hagerty duly stressed that its satellite program "will for the first time in history enable scientists throughout the world to make sustained observations in the regions beyond the earth's atmosphere."[61] America's three successful Explorer satellites did not turn out to be what Hagerty called a "unique opportunity for the advancement of science," since they shared earth orbit with an equal number of Soviet Sputniks. But they made significant contributions to IGY science just the same. The Explorers' geiger counters, designed by American geophysicist James Van Allen, discovered the two donut-shaped belts of cosmic particles held fast in outer space by earth's magnetic field. By revealing these "Van Allen Radiation Belts," the US satellites extended the range of IGY research and allowed geophysicists to begin integrating their analysis of the high atmosphere and space environments. The US

Department of Defense (DOD) attempted such an integrated analysis when it exploded three low-yield atomic bombs nearly 300 miles high. Coinciding with the IGY, but not part of it, these top secret "Project Argus" bomb tests created earth-circling shells of energetic particles. An Explorer satellite detected these particles, which triggered auroras seen by IGY observers throughout the southern hemisphere.

Information streaming out of its IGY satellite program was a boon not only to science but also to America's security. Van Allen called Project Argus "one of the greatest experiments in pure science ever conducted."[62] But the DOD designed the experiment in large part to determine if artificial radiation belts emanating from enemy nuclear explosions would blow out the electronics of planned military satellites. The IGY orbiter that played a part in that experiment also profiled the electromagnetic characteristics of the near space environment, showing engineers that future spy satellites needed to be hardened against naturally occurring radiation as well. By sharply delineating the shape of the earth and varying properties of its atmosphere, these first Explorers and subsequent satellites helped military strategists better target long-range bombers, planned Intercontinental Ballistic Missiles (ICBMs), and future submarine-launched nuclear weapons. Defense officials anticipated these military applications, but expected legal returns and enhanced national morale and prestige to be the greatest strategic payoffs of America's IGY satellite program.

At a time of resurgent Cold War confrontation, the Eisenhower administration urgently sought effective means to determine Soviet military capabilities. Its so-called New Look military strategy depended on such accurate and up-to-date reconnaissance. In his effort to limit defense spending by tilting US armed forces toward high-tech nuclear bombs and delivery systems, President Eisenhower needed to know the minimal arsenal the United States required to counterbalance its adversary and impress allies with its deterrent capabilities. He tried to get that information through diplomacy and hoped the Soviet Union would agree to an "Open Skies" policy whereby each nation allowed the other to reconnoiter its territory by air. When Soviet General Secretary Nikita Khrushchev rejected "Open Skies" as a ploy for US espionage, Eisenhower fell back on tricky automated surveillance balloons and then high-altitude U2 spy planes to fly over the Soviet Union and gather this information. The risk of these flights became plainly evident when the USSR famously shot down a U2 in May 1960 and imprisoned its pilot Francis Gary Powers for nearly two years.

Desperate for a reliable and legal method of reconnaissance, the Eisenhower Administration considered spy satellites. The RAND

Corporation, a think-tank spun off from the US Air Force, produced the first of several detailed studies of the military value of earth-orbiting satellites in 1946.[63] Worried about the possibility of a nuclear age Pearl Harbor, members of President Eisenhower's Surprise Attack Panel, given the less alarming name the Technological Capabilities Panel (TCP), echoed RAND and recommended in 1955 that the United States build a satellite large enough to hold still bulky surveillance equipment. Assuming this would take several years of R&D, the TCP proposed that a small scientific package precede a far trickier military orbiter. The NSC concurred and determined in its secret policy document NSC 5520 that "while a small scientific satellite cannot carry surveillance equipment and therefore will have no direct intelligence potential, it does represent a technological step toward the achievement of the large surveillance satellite."[64] More important, since such a satellite "would constitute no active military offensive threat to any country over which it might pass" and would therefore "test the principle of 'Freedom of Space'," the NSC endorsed a US scientific satellite program associated with the IGY. Excited about space-based scientific observation, the NAS had proposed such a program several months earlier. Now that the Eisenhower administration desperately wanted to develop the capabilities and legal precedent for orbital surveillance, the DOD's Advisory Group on Special Capabilities evaluated bids from three military services and settled on the Naval Research Laboratory's Vanguard proposal for the US IGY satellite program.

Historian Walter McDougall contends that the Advisory Group worried that other nations would associate the more promising Army bid with that service's ballistic missile program and therefore chose Vanguard because it projected a more civilian profile. With a proposed satellite designed by the NSF and NAS and launched on a Viking sounding rocket, McDougall reasoned, Vanguard might trail the Soviet Union into space but it had the best chance of conforming to IGY standards, enjoying international approval, and establishing the all-important legal principle of freedom of space.[65] Scholars have demonstrated that a divided Advisory Group in fact believed Vanguard was technically promising and could very well be the first satellite to orbit earth. Like every major study on prospective satellites to date, the Advisory Group recognized that America's soft power was at stake and hoped to shore up US prestige by winning the race into space. RAND's 1946 report presciently declared that the "achievement of a satellite craft by the United States would inflame the imagination of mankind, and would probably produce repercussions in the world comparable to the explosion of the atomic bomb."[66] Alerting Defense officials of the psychological effects of such a satellite, a veteran Manhattan Project physicist predicted in 1953 that throughout the world "the spectacle of a man-made satellite would make a profound impression on the minds of the people."[67] The Central Intelligence Agency (CIA) agreed with national security advisors on the TCP and NSC and warned that the "psychological warfare value of launching the first earth satellite makes its prompt development of great interest to the intelligence community and may make it a crucial event in sustaining the international prestige of the United States."[68]

The gathering threat posed by Soviet weapons development forced the United States to maintain the credibility of its high-tech military deterrent through overall technological leadership. As DOD analysts bleakly concluded, "Failure to maintain technological superiority by the U.S. could result in loss of confidence by the Free World in U.S. technology and power; accelerated Soviet expansion geographically and economically; swing of important uncommitted nations into the Soviet orbit; [and] defection of important countries now members of the Free World." Flanked by Moscow's diplomatic outreach to less developed countries, the United States also had to broadcast its technological superiority in ways that demonstrated "its peaceful intentions."[69] According to historian Kenneth Osgood, the Advisory Group thus endorsed what it hoped was a winning IGY satellite program to signal US technological leadership, establish the principle of space over flight, and demonstrate America's peaceful ambitions.[70] The Advisory Group approved Vanguard in particular, historian Michael Neufeld adds, because its unclassified, nonmilitary specs allowed scientific cooperation with other IGY nations, while its promising technology would boost US prestige by reaching earth orbit before a Soviet satellite.[71] Despite this careful planning, the freedom of space over flight was established not by Vanguard but by Sputnik, the victorious Soviet contender in this first leg of the space race. Sputnik also confirmed American analysts' many predictions about the psychological ramifications of orbital satellites, but it did so in an unanticipated way. The Soviet Union enjoyed the global limelight when it reached space first, while dumbstruck Americans worried that their nation's all-important prestige had been perilously diminished.

Space, Antarctica, and the Sputnik Crisis

The Cold War entered an ominous new dimension when the Soviet Union orbited its first Sputnik on October 4, 1957. America's leadership of the Free World, which depended on its superior science and technology, suddenly seemed threatened. Its scientists and engineers spawned life-enhancing products, agriculture and industry boosting technologies, and advanced weapons systems that tilted the global balance of power toward the United States. Although the USSR subsequently developed many of these technologies, Americans reassured themselves that the United States would maintain its critical lead since these wonders sprang naturally from the laboratories of free people. As one prominent journalist reassuringly wrote, "Soviet achievements in science and technology...were simply the fruits of espionage."[72] Sputnik and its two larger and technically sophisticated successors may not have shattered America's strategic advantages, as many alarmists feared, but they put into question its military superiority and challenged the common assumption that path-breaking R&D flourished only in democratic, capitalist societies like the United States. As *Time* magazine brooded, it "was becoming all too apparent Russian scientists are as good as any in the world- or better."[73] Among the many ways national leaders tried to counter this threatening impression over the ensuing decade, they projected America's peaceful intent and superior science and technology through its leading exploratory programs in outer space and Antarctica.

President Eisenhower remained unflapped by Sputnik. Even though he was surprised by the satellite's impressive size and early launch date, the president declared it had little military value and was nothing more than "one small ball in the air."[74] The powerful rockets that orbited the nearly 200 pound Sputnik and its gargantuan successors were another story altogether for they confirmed Moscow's earlier announcement that it had functioning ICBMs. Combined with Soviet hydrogen bombs, an advanced version of which test detonated just days after Sputnik, these missiles posed an existential threat to the United States. The USSR would soon be able to launch a devastating thermonuclear attack from its own soil on the distant United States. Nevertheless the president remained committed to "restrain those elements in government and society willing to jettison limited government and financial restraint in order to prove American superiority," in the words of Walter McDougall, and he refused to hastily answer the Sputniks with a budget-busting space program. Instead, he stepped up covert surveillance of Soviet military capabilities and counseled Americans to remain calm and confident in their strategic posture.[75] Reconnaissance photographs confirmed America's military advantages, but that top-secret data necessarily went unpublicized and therefore failed to temper a public debate whipped about by political leaders and national media.

Widely considered a profound moment in human affairs and the epochal opening of the space age, Sputnik naturally attracted enormous public attention in the United States. It was the satellite's blow to US military superiority and prestige that most concerned national leaders, especially the Republican president's partisan opponents, and gave the story enduring front-age attention. Administration officials quietly welcomed Sputnik as a legal precedent for freedom of orbital overflight, thereby allowing US satellites to sweep over Soviet territory, and they publicly assured Americans that their national security remained intact. Such assurance was a hard sell since America's strategic reliance on nuclear deterrence had recently failed to rollback North Korean aggression, prevent communist gains in Indochina, and stem the drift of Middle Eastern nations into the Soviet camp. When Sputnik indicated the United States might not have the leading arsenal necessary to prop up an already hard-pressed containment strategy, national headlines revealed a shaken public faith in America's high-tech military preeminence.[76] As *Life* magazine neatly summarized, it "had taken [the Soviets] only four years to break our A-bomb monopoly. It took them nine months to overtake our H-bomb. Now they are apparently ahead of us on intercontinental ballistic missiles."[77] With America's once secure airspace potentially laid bare by these world-circling weapons, Democratic politicians like Missouri Senator Stuart Symington groused that Sputnik revealed a "growing Communist superiority in the all-important missile field. If this now known superiority develops into supremacy," he predicted, "the position of the free world will be critical."[78] Senate Majority Leader Lyndon Baines Johnson opened his long-running Preparedness Subcommittee hearings on outer space in November 1957 with the similarly grave declaration: "With the launching of Sputnik I and II and with the information at hand of Russia's strength, our supremacy and even our equality has been challenged. Our goal is to find out what is to be done."[79]

Experts testifying at the hearings agreed on what needed to be done. They felt US security had suffered a grievous blow and required the hard power of leading rocket and space programs. Many of these experts were self-interested military officers who wanted to expand Army, Navy, or Air Force missile programs. Nevertheless their sobering counsel became conventional wisdom as it echoed across Washington, filled front-page newsprint, and appeared in the leaked top-secret "Gaither report" that called Soviet rockets a perilous threat to national security.[80] Democratic organizations and politicians, including party nominees for top national office, played up this terrifying challenge to US security and turned a misperceived "missile gap," in which the United States apparently trailed the Soviet Union in missile development, into a potent platform of their winning 1960 presidential campaign.[81] Despite Eisenhower's steady composure, many public officials and pundits feared that America's strategic posture had weakened owing to Sputnik's swat not only at US hard power but also at its soft power. As historian Walter McDougall points out, America's leadership of the free world rested not only on its "superiority in the technology of mass destruction, shielding those under its umbrella from external aggression" but also on the prestige generated by the related "superiority of American liberal institutions, not only in the spiritual realm of freedom, but in the material realm of prosperity."[82]

Again many Americans regarded economic prosperity and military superiority as natural offspring of a free society that provided the perfect environment for scientific research. Historian David Hollinger explains that they insisted that "science and democracy were expressions of each other and that both were threatened" during the war by fascism and after by Soviet communism.[83] In their view Nazi Germany failed to develop an atomic bomb and postwar Russia was reduced to stealing American weapons secrets because good science could not take root for long in their autocratic societies. The noted sociologist Robert Merton provided a theoretical basis for this claim by arguing that the scientific enterprise required free inquiry and open deliberation among its practitioners and was therefore the clearest and most productive exercise of democracy in the modern world.[84] NSF Director Alan Waterman put Merton's theory in compelling laymen's terms shortly before Sputnik when he explained that because "each new scientific finding, even by a Nobel prize winner, is challenged and subjected to critical evaluation and test by others," the process of science "is thoroughly democratic. In fact," he continued, "scientific research has always been, in this sense, a 'free enterprise' system."[85] Several years later a State Department official drew a link between this theory and American foreign policy by suggesting that its leading science can "enhance the prestige, leadership, and influence of the U.S." because "its possession of unity, universality and independence makes it truly supranational in character." Science was not only a powerful expression of humankind's democratic inclinations, he asserted, it was also a bulwark of universal liberal aspirations for it "reduces the tyrannical control of man's mind even in the most rigid dictatorship."[86]

The commonplace equation of advanced science and technology with superior politics and social organization, presumably those of the free and democratic United States, came home to roost after Sputnik. Ambassador Clare Booth Luce, retired congresswoman and wife of Henry Luce, pinpointed the challenge to the nation's connected hard and soft power thus:

we ourselves have made it an article of world faith that the nation which builds the biggest bombs must have the best morals, and that the most moral nation will always build the biggest bombs. We need not be surprised today that Soviet Russia is making the same claim...And we should not be surprised tomorrow if other nations believe and decide to hitch their wagon to her Beeping Star.[87]

The *New York Times* weighed in by mocking the Soviet "attempt to persuade people, especially in Asia and Africa, that Moscow has taken over world leadership in science." But *Life* magazine warned that the "Sputniks give this old Communist swindle a new lease of plausibility."[88] This swindle was taken seriously by a ranking government official who worried that because "the achievements of a nation in outer space may be construed by other nations as dramatically symbolizing national capabilities and effectiveness," the international community might swallow communist propaganda that "the Soviets ride the wave of the future."[89]

US embassy and consulate personnel wired home troubling accounts of such propaganda. Their State Department colleagues in the United States Information Agency (USIA) collected newspapers and conducted polls around the world that indicated Moscow's information campaign worked. The *Washington Post* reported in October 1960 that USIA polls indicated America's "five major friends in Europe have become increasingly convinced that the Soviet Union's space feats presage a Communist trend to be best in everything."[90] This conviction was evinced again that year when a Belgian research foundation warned of "a new and massive challenge...to Western science from the Soviet bloc" and stressed "the utmost importance to increase the effectiveness of Western science."[91] Based on many such polls and studies, the USIA concluded that the "public awareness of the first Sputnik was almost universal," that Soviet prestige has risen, and that the principal danger to US hard and soft power was the "cockiness engendered in Soviet officials themselves."[92] Such cockiness came off in official Soviet announcements. *Pravda* trumpeted Sputnik I as evidence of "how the freed and conscientious labor of the people of the new socialist society makes the most daring dreams of mankind a reality."[93] As the USSR racked up several more high profile firsts in outer space, Kremlin publicists hailed them as signs of a "new era in human progress" and the "embodiment of the genius of the Soviet people and the great might of socialism."[94]

Worried that people at home and abroad might buy this propaganda, many prominent Americans insisted their country was still the natural repository of scientific genius and achievement. They embraced the well-worn argument Vannevar Bush made years earlier that because "Russia is a closely controlled dictatorship, a police state, with full ultimate management of the details of the life of every citizen," it could produce awesome science and technology "in the short run." Bush insisted that in "the long run," however, "a totalitarian state cannot compete with a free people in the advancement of science, for the dictation and dogma are contrary to the free spirit of inquiry, which is the heart's blood of scientific advance."[95] After Sputnik a prominent media executive similarly allowed that "Russia—or any other dictatorship—has a certain head start on a democracy. One man, or a handful of unanswerable men, make all the decisions....And the people obey."[96] Vice President Richard M. Nixon agreed that "a dictator state" like the USSR "can in the short-run achieve spectacular results by concentrating its full power in any given direction." But Nixon reassured Americans that "free men in the long-run will out-plan and out-produce a slave economy."[97] While the NSF's Alan Waterman also exhorted America's natural advantages, he argued it could not passively stand by as Soviet R&D impressed the world. "Whether our primary objective as a nation is to deter our enemies, to sustain the Free World's leadership, to extend a helping hand to less favored nations, or merely maintain peace and prosperity at home," Waterman exclaimed, "the first essential is a real determination to achieve better education, better science and technology."[98]

President Eisenhower broadcast his determination that the United States achieve better science and technology when he announced in November 1957 the formation of the President's Science Advisory Committee (PSAC) to furnish him with expert counsel on critical scientific issues. The White House and Congress had the same goal in mind when they passed the 1958 National Defense Education Act (NDEA). Under the NDEA, which one scholar called "the most important federal bill related to education" in nearly a century, Washington's expenditures for education more than doubled as it subsidized student loans, teacher education, curriculum development, and graduate fellowships particularly in math, science, and engineering. To build the robust educational infrastructure a free people needed to effectively tap their natural advantages, taxpayer money flushed into American universities not only through student scholarships but also as research grants for science and engineering faculty. In the wake of Sputnik, universities accelerated their postwar growth as federal expenditures for R&D grew dramatically "from $2.7 billion to more than $15 billion between 1955 and 1965."[99]

Like many other partisan opponents of the president, Pennsylvania Representative Daniel Flood felt these high-profile efforts were not enough to demonstrate incontestably America's superior science and technology. When "the desert tribes and the coastal tribes—black, white, red, and yellow—all over the world heard about [Sputnik], all they knew was this was a public image, a public manifestation of the ascendancy and the primacy over America." Flood declared that the United States needed to do more than play catch up with the Soviets. It needed to generate "a public image of supremacy" to impress "the peoples of the world of that reality of [American] power."[100] Lloyd Berkner had this public image in mind when he advised the White House and Congress that the United States could impress those people, who "congregate around the nation and the system that succeeds in acquiring those external qualities that they find attractive," by mounting leading exploratory programs in space and Antarctica.[101] Berkner predictably emphasized space exploration since it had become a global cause célèbre after Sputnik, indeed American newsmen tagged it as the most popular news topic of 1958. He likely pointed to Antarctica because it shared many strategic virtues and symbolism with outer space.[102]

As their IGY precedents demonstrated, exploratory programs in Antarctica and outer space reinforced America's hard power and attracted the close scrutiny and continuing support of its national security establishment. These programs bolstered the nation's critical soft power by promising significant scientific discoveries and substantive international collaboration. They accommodated many small-scale researchers and became iconic examples of the technology-intensive big science programs then at the vanguard of many fields of physical science and engineering. Supporting these big science programs, America's long-haul aircraft that quickly jetted over vast Antarctic distances and rockets that blasted into space at unimaginable speeds effectively projected the sublime power of its leading technology. If this was not impressive enough, these roaring conveyances opened up other-worldly environments, piquing people's romantic awe and dramatically signaling the start of a fantastic new era. Humans had always measured the world from the starting point of their middling frames, but the new atomic and space ages extended their reach deep into the microscopic and celestial realms. Standing at what seemed the normative center of a vast and now accessible universe, they now had the power to harness the atom, master the planet, and chart the unbounded cosmos. The United States publicly committed to doing all three.

Unlike its atomic energy programs, weighted by their association with nuclear weapons, America's ensuing civilian endeavors in outer space and Antarctica appeared primarily peaceful. Because they were spectacular efforts that required the focused energies of this most technically endowed superpower, many people even touted them as means to ratchet down the Cold War. They were technology-intensive programs of scientific exploration, conducted openly, amid many gestures for cooperation with the Soviet Union, and in the declared interests of all humankind. These programs of exploration evoked a favorable national myth of discovery perfectly suited to this postcolonial age. The New Worlds of Antarctica and outer space—supranational, uninhabited, and rich in resources and hidden truths of universal human value—would do for the rising global civilization what Christopher Columbus's discoveries eventually did for the free and prosperous United States. Here were dramatic opportunities to shore up US prestige, Berkner understood, for it could deploy its stunning science and technology in these vacant realms in ways that could serve and impress people the world over.

Antarctic research was dramatic and certainly served the international community, but it did not attract the intense scrutiny or scale of public investment in space exploration. The surprise of Sputnik and then subsequent Soviet satellites and piloted space flights had what many scholars call a Pearl Harbor effect on Americans. Just as they had when Japan attacked out of the blue, shocked Americans felt vulnerable and their political leaders responded forcefully, this time to reclaim US credibility and prestige through a leading space program. A January 1958 staff study of the National Advisory Committee for Aeronautics (NACA) plainly illustrated this dynamic by calling Soviet space operations the front "of a far-reaching plan and sustained effort that poses a most serious challenge to the United States and the Western world. It is of great urgency and importance to our country both from consideration of our prestige as a nation as well as military necessity," NACA advised, "that this challenge be met by an energetic program of R&D for the conquest of space."[103] The need for an energetic space program became the consensus position circulated across the media landscape and forwarded by experts to the White House. The President's Science Advisory Committee (PSAC) made its counsel public in its widely read March 1958 report *Introduction to Outer Space*. Hailed by President Eisenhower as a "sober, realistic presentation prepared by leading scientists," the report called for an affordable yet ambitious "national program for space science and technology" that would "enhance the prestige of the United States among the peoples of the world." Although national prestige was just one of PSAC's four justifications for such a program, the other three fed into this critical goal. If these advisors correctly identified a "compelling urge of man to explore and to discover" as the second justification, then a leading space program would secure America's repute for enabling humankind to follow this urge and "go where no one has gone before." Military power was PSAC's third rationale, but that too helped boost America's positive standing as the defender of the free world. PSAC's final stated purpose, "afford[ing] new opportunities for scientific observation and experiment," promised to amplify US prestige by demonstrating its unrivaled efforts to expand humankind's "understanding of the earth, the solar system, and the universe."[104]

These goals appeared repeatedly in subsequent space policy documents. They facilitated quick and productive cooperation between Congress, which passed the "National Aeronautics and Space Act of 1958," and the President, who signed the act on July 29. The resulting National Aeronautics and Space Administration (NASA) came into being with a mandate for making the United States "a leader in aeronautical and space science and technology."[105] Over the next decade it did so by erecting a multibillion dollar R&D infrastructure spread from coast to coast. Spun out of the decades-old National Advisory Committee for Aeronautics, NASA inherited that organization's five research, flight test, and rocket sounding facilities. It absorbed the Army Ballistic Missile Agency's heavy booster program, turning its Alabama complex into the George C. Marshall Space Flight Center. NASA partnered with Caltech's Jet Propulsion Laboratory, which developed expertise in robotic exploration of the moon and neighboring planets. In addition to building the Goddard Space Flight Center outside of Washington, DC and the enormous Manned Spacecraft Center in Houston, Texas, the agency assembled smaller facilities in Massachusetts, Louisiana, and Mississippi, and it occupied a large rocket-launching complex at Florida's Cape Canaveral. This arc of aerospace facilities stretched primarily across the country's southern flank, accelerated the industrialization of America's South and West, and secured the good will of millions of citizens and scores of politicians who appreciated NASA's contributions to regional economic growth.[106]

The nascent space agency prepared a long-range plan to make full use of this continental infrastructure not only for science and orbital applications but also for "manned exploration of the moon and nearby planets."[107] President Eisenhower wanted NASA to focus on science and orbital applications, so he rejected lunar and planetary expeditions as the costly and impractical scheme of people he derisively called "space cadets." NASA's first Administrator T. Keith Glennan shared the president's practical priorities and put aside a "rosy, pie-in-the-sky picture of manned space transports, civilian colonies and manned military bases on the moon or other planets" for a "more sober, commonsense picture."[108] During his short term, Glennan's picture of NASA's space operations included orbiting scientific satellites that examined earth's upper atmosphere, the size and shape of the planet, and the properties of the near space environment. The first lunar and planetary probes took shape during his tenure, as did communications and weather satellites. In these first few years, NASA experimented with inflatable orbiters that passively reflected electronic signals back to earth, contracted private industry to build active satellites to repeat telecommunications from low earth and geosynchronous orbits, and developed the rockets that would later launch satellites built for the American and international telecommunications consortiums COMSAT and INTELSAT. Although the IGY holdover Vanguard 2, which NASA inherited from the Naval Research Laboratory, failed in its mission to provide useful meteorological data, the space agency helped develop and launch the world's first functional weather satellite TIROS 1 in 1960. NASA's initial foray into piloted space travel, Project Mercury, was undoubtedly the highest profile element of Eisenhower's constrained space program. Formally approved just days after the agency opened shop, this program carried the heavy price tag of $1.5 billion and lifted six of the seven celebrity Mercury astronauts into outer space.

Those astronauts had not yet left the ground when a retiring President Eisenhower famously alerted Americans to protect their precious finances and democratic traditions from a secretive and swelling "military-industrial complex."[109] NASA's manned space operations were ostensibly civilian and transparent. But they were costly programs closely tied to that complex, liable to suck up taxpayer money in excess of what Eisenhower deemed their limited practical value. Although John F. Kennedy campaigned for muscular defense and generous government support for visionary R&D, as president-elect he heeded a transition team that similarly rejected "a crash program aimed at placing a man into an orbit at the earliest possible time." These advisors warned that such an effort "may hinder the development of our scientific and technical program, even future manned space program by diverting manpower, vehicles and funds."[110] Just a few months into his presidency, however, Kennedy dramatically changed his position. In one of the most famous presidential addresses, his May 1961 special message on "Urgent National Needs" before a joint session of Congress, Kennedy asked the nation to "commit itself to achieving the goal, before this decade is out, of landing a man on the moon and returning him safely to earth."[111] When Congress responded positively to his abrupt turnaround, NASA's budget for manned space operations went through the roof. In just eight years it spent tens of billions of dollars training astronauts and developing novel hardware for this epic undertaking, including miniaturized electronics, advanced alloys, synthetic fibers, and heat-ablating ceramics, even pens capable of writing in zero gravity and pathogen-free squeeze-tube meals. It built a global network of tracking stations to follow and communicate with its cosmic voyagers. Most dramatic of all, the space agency and its army of industrial contractors designed the ever-larger rockets that launched Project Mercury's one-man orbiters and subsequently lifted two-man capsules for Project Gemini. Finally, NASA's Saturn V rockets, the most powerful launch vehicles of all time, blasted Project Apollo's three-man spaceships to the moon, where 12 Americans put to rest any lingering questions about US power and vitality raised by Sputnik. Between July 1969 and December 1972, six teams of Apollo astronauts landed on the moon and broadcast live their age-defining footsteps over television and radio hundreds of millions of people via America's new communications satellites.

Before they settled into familiar routine, these spectacular missions accomplished their primary goal of boosting domestic pride and impressing the international community. For all their talk about Apollo's contributions to science, the US economy, and international amity, space policymakers agreed that America's manned space program was critical for national prestige. Again the country's science and technology not only reflected the exportable hard power of its military forces and vibrant economy, they also lent weight to Americans' profession of the superiority of their liberal democracy. Since the Soviet Union's path-breaking orbital operations cut to the heart of this belief, national leaders highlighted the importance of the space program to America's soft power. Thus a ranking State Department officer told Congress that because "the achievements of a nation in outer space may be construed by other nations as dramatically symbolizing national capabilities and effectiveness...our own friends are watching our future progress and achievements in this field."[112] So too did the ever-present Lloyd Berkner who insisted that "Men everywhere see in the conquest of space the peaceful demonstration of the superiority of one of the two competing systems of economic organization- capitalism versus communism."[113] T. Keith Glennan promoted a tight and efficient space program as the best, most affordable means "to demonstrate once again that free men—when challenged—can rise to the heights and overcome the lead of those who build on the basis of subjugation."[114] President Kennedy agreed that the United States needed a winning space program, he believed a winning manned spaceflight program, to enhance its standing in the world.

As many historians have documented, the president seriously considered a lunar program after the United States suffered two humiliations on the world stage in quick succession.[115] The first occurred on April 12, 1961 when Soviet cosmonaut Yuri Gagarin became the first person to orbit the planet. Then the very next week, as Kremlin publicists hailed Gagarin as the "Pioneer of the Universe" who "Ushers in a New Era of Human Progress," American-sponsored counterrevolutionaries miserably failed to storm Cuba's "Bay of Pigs" and topple the government of Soviet-leaning Fidel Castro.[116] Seeking to restore his diminished political capital and the luster of America's world leadership, the brand new president favorably received Vice President Lyndon Johnson's counsel that the United States had a good chance of winning a race to the moon. "Dramatic achievements in space" such as manned lunar landings, the new heads of NASA and the DOD had advised Johnson, would "symbolize the technological power and organizing capacity of the nation."[117] Thus when Kennedy envisioned landing men on the moon, he stated the obvious: "No single space project in this period will be more impressive to mankind" or useful to America's effort "to win the battle that is now going on around the world between freedom and tyranny."[118]

As the phenomenal cost of Kennedy's vision became apparent, a congressional report conceded that to some people "the highly advertised race to the Moon may seem merely to be a somewhat immature exercise in scientific muscle flexing by the United States and the U.S.S.R."[119] Judged by their voluminous correspondence with NASA, fevered celebration of its astronaut corps, and mass pilgrimage to view rocket launches, many citizens nevertheless embraced manned space exploration as a worthy source of national pride and prestige. In the ensuing years the State Department determined that people overseas were similarly impressed. Consular dispatches poured into Washington about local newspapers, such as one in Brazil, which hailed successful space missions as signs of the "superiority of the United States over the Russians in the development of space flight" and as a "heartening comfort for the destiny of civilization and the predominance of democracy over totalitarian powers."[120] Similar congratulatory messages from foreign heads of state flooded the White House. A South Korean leader praised an early Mercury mission as "another victory gained by the free world and the human race as well."[121] As NASA methodically trailed and then surpassed Soviet rocket power and orbital operations, USIA staff concluded in 1964 that for the first time "U.S. space feats and resulting space lead dominated reaction in the free world."[122] This conclusion rested on their many polls and reviews of laudatory headlines, including occasional stories in otherwise hostile foreign newspapers. In Lebanon for instance, the short-lived Palestinian newspaper *ash Shaab* commemorated three astronauts who died in a tragic Apollo capsule fire in 1967 as "the cream of the American people" and expressed its sorrow "for the terrible catastrophe."[123] USIA staffers and NASA public affairs officers recognized that such media largely reflected the opinions of foreign elites, both friend and foe. Thus they were heartened when cheering crowds greeted astronauts, sometimes to the consternation of foreign leaders, when they toured the world as celebrity envoys of the United States.[124]

Strategically designed to visit friends and potential allies, these tours were just one element of NASA's enormous outreach. The space agency curried public favor and quenched people's voracious thirst for information by working closely with American and foreign journalists. NASA's public affairs office mailed them press releases and technical mission briefings, shared pictures and film footage, and made sure television broadcasters enjoyed equal access to newsworthy tape of ascending rockets and astronauts in space. It hosted hundreds of reporters and VIPs who descended on Florida for every major rocket launch, responded monthly to thousands of people's letters, and fielded a dizzying number of requests for appearances by astronauts and NASA personnel, who delivered nearly 3,000 speeches in 1965 alone.[125] The space agency set up exhibits at foreign expositions and World's Fairs in Seattle (1962) and New York (1964), opened several of its centers to tourists in the late 1960s, and worked with advertisers eager to link their products to the heroic aura of space exploration.[126] NASA's Office of Technical Information and Educational Programs produced yet more exhibits, documentaries, television and radio spots, and educational curriculum.[127] Its inspired speakers and informational material reached schools throughout the United States and in some foreign countries via a fleet of distinctive vehicles NASA dubbed its "spacemobiles."[128]

Foreign outreach conducted by NASA as well as by consular officers, the Voice of America, and the USIA reached millions of people. The USIA tried to excite keen world interest by publicizing instances in which foreign nationals worked with the US space agency, as it did when it wrote a story about a Haitian employee of NASA to highlight United States-Caribbean cooperation.[129] Regardless of its unusual Congressional mandate to work with international partners, NASA reached out to many countries because it required a worldwide network of stations to track its satellites and communicate with piloted spacecraft. NASA built nearly 30 stations in more than a dozen countries. These included Mexico and the African nations of Nigeria and Zanzibar, where political leaders occasionally criticized American foreign policy. However people even there expressed great pride in their participation in a grand undertaking, in the words of NASA's founding legislation, "for the benefit of all mankind."

When these three countries balked at allowing the United States to use their tracking stations for military operations in space, they exposed a fundamental limitation in the space program's prestige value. America's standing benefited from patently peaceful displays of its leading science and technology. But its reputation and world power also depended on the credibility of its military strength. In this vein, embassy personnel sampled local opinion about the US space program in Tehran and advised State Department superiors that "military power is of the greatest importance in the building of political prestige in Iran."[130] PSAC had listed national defense as a primary goal for US space operations, as did most politicians and editorialists, one of whom neatly summarized the prevailing wisdom that the United States "cannot afford to run the risk of being out-classed by scientific breakthroughs that may make our earth-bound systems of offense and defense obsolete."[131] Washington dealt with this conundrum by separating NASA's well-publicized civilian opera-tions from the DOD's secret operations in outer space.[132] Most high-level government officials did not even know about the DOD's Corona spy satellites, which dropped enormously valuable canisters of surveillance film to earth between 1960 and 1972.[133] Many more people knew about the Air Force's rocket planes that soared to the edge of space, experimental delta-winged "Dyna-Soar" spaceship, and proposed Manned Orbiting Laboratory (MOL) that President Johnson publicly approved in August 1965. It was the possibility that their tracking stations would serve the MOL that set leaders in Mexico, Nigeria, and Zanzibar on edge, while UN diplomats wor-ried the orbital military platform would "reverse and halt the positive trend that has developed in...the peaceful uses of outer space."[134] International scrutiny sometimes fell on the civilian space agency as well. Although NASA was banned from military operations, Deputy Administrator Robert Seamans did not reveal states secrets when he declared "there is an important interchange of components and vehi-cles between the NASA and the Department of Defense."[135] His later admission that "there was a DOD spinoff from nearly everything that we were doing" became a well-worn charge made by the Soviet press, which accused the United States of militarizing space even when it launched patently peaceful weather and communications satellites.[136]

Despite the military applications of NASA programs and Moscow's regular invectives aside, the United States effectively projected its peaceful intentions by reaching out to the Soviet Union and working with the United Nations to prevent the militarization of outer space.[137] In an address to the UN General Assembly in September 1963, President Kennedy invited the USSR to work with the United States on a joint lunar landing. His foreign policy advisors rightly predicted Kremlin officials would reject this olive branch, but they felt that ensuing international acclaim warranted a second offer, which America's UN ambassador made just days after Kennedy's assassination.[138] By then the United States had worked for almost seven years with ad hoc and then permanent UN committees on outer space. Under the auspices of the United Nations, Washington took a leading role in negotiating a series of international agreements for peaceful cooperation related to space activities.[139] The most notable was the 1967 Outer Space Treaty, which extended international law into space, kept the moon and other celestial bodies free of national claims and weapons of mass destruction, and declared that the "exploration and use of outer space, including the moon and other celestial bodies, shall be carried out for the benefit and in the interests of all countries, irrespective of their degree of economic or scientific development, and shall be the province of all mankind."[140]

These efforts suited US interests and projected an image favored in Washington of a confident and powerful nation benevolently plying its way into space for all humankind. When America's nearly decade-long effort to reach the moon paid off, US foreign-service officers indicated that image-making campaign had worked. Having observed enthusiasm reactions in North Africa to the first lunar landing on July 21, 1969, they wired back that:

> soft pedaling of nationalistic sentiments in our public statements— and the avoidance of claiming any "victory" over the Russians—have succeeded in establishing America's technological supremacy and in adding to our overall national prestige much more effectively than a "hard sell" ever could have.[141]

Apollo 11 commander Neil Armstrong soft-pedaled as well. Upon becoming the first person to set foot on the moon, he put aside narrow nationalism in his now famous words: "That's one small step for man, one giant leap for mankind." Looking for a phrase at that historic moment to best express America's enlightened internationalist approach to space exploration, the White House and NASA approved similar

language for that mission's commemorative plaque. Left behind on the lunar surface, it concluded: "We Came in Peace for All Mankind."
     Widely heralded as a turning point in human history, the lunar landing outshined other national endeavors. Indeed Antarctica attracted far less public interest than outer space, but it too figured prominently in policymakers' efforts to bolster US interests during and after the IGY. The imprint of that global research program remained especially strong in Antarctica as America's continuing operations there helped compensate for Sputnik's blow to US power and prestige. Although the IGY gave the Soviet Union a permanent foothold in Antarctica, it sublimated territorial conflict and established a precedent for ongoing cooperation. Secretary of State John Foster Dulles futilely wanted to exclude the USSR from the area because he believed its IGY installations "could potentially develop into air and submarine bases from which to dominate much of the Southern Hemisphere."[142] A JCF member gave voice to another strategic concern and warned "it would create doubt in the minds of our free-world military partners as to whether we were really serious about our endeavor to contain and isolate the Russians were we to invite them into an international regime for Antarctica."[143] The United States nevertheless had little room to maneuver, especially after Soviet representatives announced in late 1957 the USSR would stay in Antarctica after the IGY. State Department realists therefore endorsed such a regime so as to coopt rather than roll back Soviet activity there. In May 1958 President Eisenhower invited the IGY countries active in Antarctica to negotiate a treaty to keep it "open to all nations to conduct scientific or other peaceful activities."[144] According to a secret State Department report, such a treaty was America's best diplomatic option. Seeking above all to maintain "a position of leadership in Antarctica for political, including prestige, purposes," it concluded that the United States should claim territory only if a "treaty among the countries having a direct and substantial interest in Antarctica, including the USSR" failed to "lesson the possibility of Antarctica becoming the scene of international discord."[145]
     A US territorial claim proved unnecessary since successful negotiations led to the 1959 Antarctic Treaty, which made IGY-style research the centerpiece of an exclusive multinational political regime. In this model for the future Outer Space Treaty, 12 signatories agreed to keep Antarctica free of weapons and nuclear materials, hold territorial claims there in abeyance, and allow each to inspect all stations and travel freely throughout the continent. Most important, the Treaty hailed its IGY roots as it explicitly encouraged "international

cooperation in scientific investigation" and made such research the qualifying criterion for full political participation in Antarctica. To become an empowered Contracting Party to the Antarctic Treaty, a nation had to "demonstrat[e] its interest in Antarctica by conducting substantial scientific research activity there, such as the establishment of a scientific station or the despatch [sic] of a scientific expedition."[146] In effect, the Treaty elevated science over traditions of discovery, settlement, and force of arms as the legal basis for states to assert political influence over Antarctic affairs.[147]

American critics of the Antarctic Treaty warned that Moscow intended to dominate the world's icy white underbelly. An outraged Senator said it "amounts to putting the free world and the slave world on the same footing," and a hawkish Congressman called it a "humiliating" document that would be "misinterpreted by the Soviets as evidence of our weakness and our willingness to appease and retreat."[148] The anticommunist National Sojourners similarly pilloried the Treaty and called on President Eisenhower to make a "formal claim without delay to the regions of the Antarctic to which the United States has rights by reason of discovery and exploration."[149] So too did an agitated citizen who pleaded with her senator: "in the name of common sense (if not in the name of God) let our leaders in Washington file a claim in Antarctica."[150]

Conventional rhetoric about a global threat of communist slavery aside, the United States had little option but to work with its Soviet adversary in Antarctica. When the president signed and Congress ratified the Antarctic Treaty, they turned America's limited hand into an important foreign policy coup. They forestalled Soviet territorial claims in Antarctica, protected America's access to the whole continent, and signaled its readiness to work alongside other nations in the region for the declared benefit of humankind. As the State Department predicted, this was a boon for US prestige. People the world over saw that the United States truly embraced international cooperation, and they could find comfort that its diplomatic initiative resulted in the demilitarization of a whole continent and a precedent for further military inspections and arms control agreements. As *Newsweek* magazine cheerily opined, this represented "a chance, a slim chance but a real one that a start may be made on ending the Cold War in the coldest place of all- Antarctica."[151]

The United States appeared to put aside Cold War concerns and focus on its unsurpassed scientific program in Antarctica. Advised by a committee of polar specialists of the NAS, the NSF managed the scientific aspects of the post-IGY United States Antarctic Program (USAP).

The NSF dispensed several million dollars a year in grants to university scientists to carry on Antarctic research that came of age during the IGY. Glaciology, meteorology, and upper-atmosphere physics remained key fields of science in Antarctica that advanced basic knowledge and filled a huge geographical gap in ever more sophisticated models of earth's physical environment. Astrophysics continued to be a staple of Antarctic research as scientists turned the high-altitude ice sheet into an exceptional platform to view the cosmos and assess the electromagnetic properties of near space. The USAP also encouraged science not officially sanctioned during the IGY. Then considered politically caustic to the delicate diplomatic dance then occurring in Antarctica, cartography and geology became standard fields of research after the IGY. Whereas IGY parties feared accurate maps and geological surveys might antagonize claimant nations, the Antarctic Treaty parties welcomed them as valuable contributions to their scientific programs. Zoology, ecology, and the biomedical sciences became standards as well. Specialists in these fields visited Antarctica during the IGY; thereafter they became permanent fixtures in America's scientific bases across the continent.

Logistical support for American research initially cost tens and later hundreds of millions of dollars a year. That support flowed through the USAP operational center at Antarctica's ice-bound McMurdo Sound, which even housed a small AEC nuclear power plant for ten years to heat, electrify, and desalinate water for the complex. US Coast Guard icebreakers annually cut a path to that base for Navy and merchant marine supply ships. Overland conveys embarked from McMurdo, which still hosts the ice shelf airport from which most helicopters, light aircraft, and Air Force and Army transport planes take off for destinations around the continent.[152] In addition to temporary field camps and McMurdo, the United States refurbished three of its other six IGY stations. Working at the extreme end of America's global logistics network, meteorologists, upper atmosphere researchers, and astrophysicists thrived 850 miles away from McMurdo at the South Pole Station, while these scientists as well as glaciologists specializing in ice core analysis hunkered down at the remote Byrd Station. Biologists found a cramped home for several years at Hallett Station and then at the slightly more spacious Palmer Base built along the coveted Antarctic Peninsula in 1968. Zoologists and oceanographers traveled to these two coastal stations and worked from the decks of a several increasingly sophisticated research vessels the NSF kept afloat around Antarctica.

Owing to the USAP's operational needs and foreign policy import, the NSF shared primary governance with ranking officials in the

Departments of Defense and State. Representatives from these three agencies sat on the short-lived Antarctic Working Group of the NSC's Operations Coordinating Board, subsequently worked together ad hoc for several years, and then led the Antarctic Policy Group (APG) called into life by President Johnson in April 1965. The committee's focus on diplomacy is evident in the leading role played by the State Department and the APG's goal of "foster[ing] international cooperation among the nations active in Antarctica."[153] That cooperation sometimes took the form of US logistical assistance for its many Antarctic partners, but it usually entailed facilitating scientific collaboration. IGY-style collaboration continued formally in 1959 during a follow-up year called the International Geophysical Cooperation, during the 1964 International Year of the Quiet Sun, and in regular multinational research projects organized through ICSU's Scientific Committee on Antarctic Research. American oceanographers plying Antarctic waters worked with international partners through ICSUs Scientific Committee on Oceanographic Research, and US meteorologists fed their observational data to the World Meteorological Organization. The United States handed two of its IGY research bases to other nations, shared its Hallett Station with New Zealand, and quite remarkably swapped scientists with the Soviet Union in every year but 1959 for seasonal stints at each others' Antarctic bases.[154]

As they promoted American research and scientific collaboration, government officials also promised that humankind would enjoy substantial payoffs from that activity. A top figure in the State Department modestly suggested that "no one can foresee what in 10 years might be the utility or the value or the usefulness to mankind of that area."[155] But the NSC was less ambiguous as it repeatedly said "determining the nature and extent of resources" was one of America's top goals in Antarctica.[156] Proponents of McMurdo's experimental atomic power plant predicted it would help open the continent for comprehensive economic exploitation, while other government and media publicists more narrowly envisioned radio repeaters and satellite tracking stations in Antarctica, cold weather storage for surplus crops, mineral and marine resource extraction, regular commercial air routes overhead, and even nuclear waste disposal under its thick ice sheets.

Public officials celebrated science and speculated openly about the economic benefits of the USAP, but they kept very quiet about US strategic maneuvering in Antarctica. Because Americans and their friends and potential allies wanted to know that the United States had a preeminent defense posture around the world, these maneuvers also reinforced national pride and prestige. But they could not be openly acknowledged without casting a shadow over America's vaunted peaceful conduct in the region. NSF Director Alan Waterman therefore confessed in confidence that because "the level of Antarctic activity cannot properly be determined nor justified on scientific or technical grounds alone," military and foreign policy objectives were foremost in the minds of top Antarctic policy makers.[157] Representatives from the Departments of Defense and State had front-row seats on the Antarctic Policy Group and they clearly steered US Antarctic affairs. They did not advertise, however, that naval commanders initially regarded their Antarctic operations as good preparation for maneuvers in the Arctic, which many military strategists identified as a potential cold war battlefield. Antarctic operations improved Navy psychiatric screening and heightened its experience in polar construction, communications, and travel.[158] Nor did policymakers publicize CIA assessment of Soviet capabilities and that spy agency's recommendations for minimizing its concomitant military threat in Antarctica.[159]

State Department officials quietly endeavored to make America's strategic position on the icy continent superior to that of the Soviet Union. They approved costly logistical support for Belgium's base in 1960 and 1963 so as to use that ally's station to offset Russia's unrivalled presence in eastern Antarctica.[160] They advised the NSF as to which research activities would lay the legal basis for American territorial claims in the event the treaty system fell apart.[161] Foreign policy specialists wanted to establish a permanent station on the hotly contested Antarctic Peninsula, and they favored ambitious mapping expeditions and icesheet traverses near Soviet bases, including one in 1967 that "projected U.S. interests into what they fear is becoming known as the 'Soviet Sector' of Antarctica."[162] Even the Antarctic Treaty's celebrated system of open inspections of scientific bases was subject to strategic calculations. Jittery State Department officials, for instance, did not want to invite Chile to inspect US bases in Antarctica. They assumed the Soviets might respond by examining that country's south polar stations, thereby enraging Chilean anti-communists who would demand their country abrogate the Antarctic Treaty and assert its territorial claim.[163]

Policymakers kept these particular objectives under wraps since the United States garnered critical prestige from activities that were overtly peaceful, valuable to all people, and more extensive than those of its Antarctic partners. Although it required the hard power of credible military force, America's world leadership depended on the soft power of its positive reputation for peaceful internationalism. Thus when the

Navy produced a film depicting western "security through sea power" around the southern continent, an NSF official worried it would cast an unfavorable shadow over US Antarctic activities and diminish what his boss Alan Waterman deemed "a real asset in prestige."[164] The State Department quietly cultivated this asset from the outset. It was solicitous of postcolonial nations, including those already disturbed by racist South Africa's role in treaty negotiations, and secretly warned that if "Japan is not invited" to enter the Antarctic Treaty, "it will have no Asian representation and be a strictly white man's club."[165] Japan solved that problem by adding some color to the Treaty. US officials focused their attention of course on the Soviet Union. Worried that it wanted "to demonstrate superior Communist capabilities by their achievements in Antarctica," they privately urged the United States to "maintain a position of leadership in Antarctica for political, including prestige, purposes."[166] The CIA offered the same counsel, as did presidential science advisors who recommended Antarctic activities "adequate to match in prestige and leadership the contemplated Soviet program."[167] Those activities included a full spectrum US research program, keeping its remote South Pole station operating year-round, maintaining an enduring presence around the continent, and building the most advanced air transport system in Antarctica. While Russian tractors laboriously crawled across select regions of crevassed ice, the United States quickly jetted anywhere in the continent. State Department officials even reversed their standing policy against supporting private expeditions in Antarctica. They justified naval transport for a politically connected group that wished to scale its highest mountain by arguing that "the U.S. will earn prestige if it is the first to climb the peak."[168]

Government agencies most involved became active publicists for the USAP. Hoping to curry favor for their programs by enhancing what they called "public understanding of science," NSF officials had already ramped up media outreach to impress Americans with the importance "of science and its relationship to their daily lives and the future of the Nation."[169] Once the agency took responsibility for US Antarctic research, its public affairs officers reached out to influential reporters, politicians, and businessmen so that they could help spread the gospel of the USAP. The NSF outfitted these "distinguished" visitors at considerable expense and flew them to Antarctica, where it showed them first hand how the United States sought new knowledge, natural resources, and opportunities for international cooperation on that icy continent. Still aglow from their south polar adventures, returning congressmen praised the NSF for the "great job that is being done in the interest of all of us," while journalists detailed the heroic contributions Americans made to global peace and prosperity on the barren continent.[170] Filmed during such a visit, a 1960 *Science News Digest* movie celebrated in this vein "the march of machines and men from many nations, mobilized in an unprecedented example of international cooperation," envoys of a world civilization who have "come to Antarctica to stay, carrying the light of knowledge to the very ends of the earth."[171] NSF staffers favorably reviewed a filmmaker's proposal two years later for a movie that would show in America and serve as a "peace tool in the hands of the United States Information Agency," projecting "to the rest of the world...a courageous, enthusiastic, vigorous 'image' of the modern American scientist."[172] The filmmaker reasonably assumed USIA interest since it and the Voice of America had broadcast similarly upbeat stories around the world about American research and international cooperation in Antarctica.

The highest-level boosterism of this sort came in the White House annual message to scientists wintering over in Antarctica. Signed by the president but carefully scripted by his advisors and specialists in the NSF and Departments of State and Defense, these so-called midwinter messages reflected an official consensus to publicly emphasize international cooperation on the southern continent.[173] Thus in language characteristic of chief executive statements about space exploration, President Eisenhower saluted those researchers "in the name of science for the work you are performing and the sacrifices you are undergoing for the benefit of all mankind.[174] President Johnson did the same. Declaring that the "selfless dedication of people from many nations to the advancement of man's knowledge symbolizes the aspirations of mankind everywhere," he applauded "the harmony and friendship which characterize your relationships in Antarctica [as] an example of international trust and cooperation for the world to emulate."[175] Finally by rhetorically tying Antarctica to outer space, both presidents linked two programs of scientific exploration launched during the IGY and accelerated in varying degrees by Sputnik. Speaking to the United Nations General Assembly, Eisenhower proudly pointed to the Antarctic Treaty as a model for international cooperation in outer space.[176] Johnson chose an even loftier comparison and said the "challenge in Antarctica is similar to the challenge in outer space." Both regions may change people's "fundamental conceptions of nature," he intoned, as they "challeng[e] the ingenuity of scientists, engineers, and explorers in overcoming obstacles of nature which have frustrated and baffled the human race through the centuries."[177]

Conclusion

Presidents Eisenhower and Johnson identified a common global mission for US space and Antarctic exploration. If they divined world peace and prosperity as that lofty mission, Lloyd Berkner cut straight to the fundamental national interests driving these federal programs. The United States invested in them to buoy domestic pride, reinforce the faltering credibility of America's superior military and economic power, and enhance the nation's prestige as a laudable world leader demonstrably inclined to peaceful internationalism. Such Cold War calculus was evident as early as the IGY, when Washington initiated these programs and much of the world took for granted America's scientific and technological leadership. That leadership aided its goal of leading a secure and economically integrated world, and it made the liberal dream of global peace and prosperity seem more likely than the millennial vision promoted by Soviet communists. Had the United States fulfilled world expectations and orbited a satellite first, it may have quietly soldiered on in Antarctica and outer space after the IGY. But Sputnik upset the Cold War balances of hard and soft power and thrust spaceflight and to a lesser extent the US Antarctic Program into the public eye at home and abroad.

As these technology-intensive big science programs grew over the course of the ensuing decade, institutional promoters treated them as fitting answers to the Soviet Sputniks. They also looked to make these programs meaningful and secure long-term public support for them. They did so by wrapping them in the potent American story of pioneering nationhood. That mythic national identity exalted a special people who used their unique dispensation of the vacant New World and western frontiers to nurture the liberty and prosperity craved by all people. Now that the United States lacked such hinterlands, just as it needed a pioneering citizenry to face its Cold War challenges, boosters and media pundits cast American astronauts and Antarctic explorers as the frontier vanguard for freedom-loving people the world over. In this way outer space and Antarctica became frontiers for the American Century.

Chapter 2

The Space and Antarctic Frontiers

In the decade after the International Geophysical Year (1957–58), the United States established a leading presence in outer space and Antarctica. Had America been just a regional power, it would not have undertaken extensive programs of exploration in these distant realms. Such countries did not have the resources to do so, and their relatively modest standing in the world did not hinge on such high-profile endeavors. The United States, however, was a superpower whose aim to contain the Soviet Union and lead a global bloc of "Free World" nations was well served by its space and Antarctic programs. Satellites aided America's worldwide military posture with reconnaissance and communications, and its activities in Antarctica helped keep the peace there and gave it regional influence in case that amity broke down. US space operations sparked industrial innovation and spawned a telecommunications network that closely linked the international community, while many people pointed to Antarctic resources as an imminent boon to the global economy. Most important, the country's dramatic efforts in these forbidding realms enhanced the domestic pride and international prestige necessary for world leadership and gave the United States a visionary luster commensurate with the lofty goals of the American Century.

Shortly after Henry Luce shared his dream of an American Century in 1941, his countrymen followed his cue and fought against fascist empires in Europe and Asia. Having internalized national myths of circled wagons and embattled forts, many saw their country besieged by savage enemies once again after the war and accepted unprecedented peacetime internationalism to contain a communist menace. A fortress mentality may have mobilized Americans and attracted foreign allies who dreaded the Soviet Union and their own homegrown radical movements. But fear alone could not sustain America's leadership of the Free World. Such a global endeavor had to be more than a rearguard action against an offensive foe. It needed to be a hegemonic project with forward-looking goals that appealed to people at home and abroad. That was Luce's vision of the American Century, a new age in which the United States inspired humankind to achieve collective freedom and prosperity through political and economic modernization. Those goals animated American cold warriors as appealing alternatives to Soviet propaganda and to communism, which attracted so many people in war-torn and newly independent countries. They averred that democratic capitalism could liberate nations and bind them together since it was perfectly attuned to human aspirations and most effective in tapping people's innate talents. The country's unrivaled production and per capita wealth seemed to confirm its ideological superiority, as did path-breaking science and technology associated with America's leading space and Antarctic programs. These metrics alone, however, illustrated raw US power rather than the high purpose on which national pride and prestige also depended. Advocates and media observers evoked that purpose by wrapping these programs in nationalist myths familiar to Americans and well suited to their internationalist agenda. Mythic tropes of the United States as a beacon of freedom and a New World model of progress appeared in public discourse, but the "frontier" became the most common motif for US space and Antarctic exploration.

This was especially the case with outer space, which popular entertainment had long filled with the types of heroes and pageants projected onto America's western frontiers. At the dawn of the Space Age many public authorities followed suit and described the operations of the National Aeronautics and Space Administration (NASA) as the first fledgling steps onto a cosmic frontier. Now that it served their institutional interests, they turned a fanciful storyline into a sober selling point used to encourage national support for space exploration. Since the US Antarctic Program (USAP) was far less costly and depended on fewer interest groups, its advocates had less need of a compelling trope to market their favored program. But they too spoke of Antarctica in these terms, calling it the world's last geographic unknown and a steppingstone to the final frontier of outer space. Historians have overlooked this application of the frontier motif to Antarctica, and those who recognized its relevance to outer space have largely detailed what appeared to be unchanging rhetoric of the space frontier. Perhaps scholars rarely plumbed the motif's deeper meanings and people's motivations for using it because they assumed that Americans reflexively adopted a familiar, monolithic, and timeless trope. But proponents reached for a multifaceted frontier motif because its timely lineaments, which later fell out of synch with national priorities and sensibilities, helped them market space and Antarctica during the late 1950s and 1960s as frontiers for the American Century. This most potent of nationalist myths ennobled their mundane interests in NASA and the USAP and enabled media producers to spin compelling stories of grand national action at the ends of the earth. It also excited many Americans who nostalgically yearned for geographic frontiers. The space and Antarctic frontiers promised new outlets for their creative energies and spoke to their hope that a vital United States would forever be the world's foremost power. Thus this frontier motif was not simply a branding strategy conjured up by program salesmen. It reflected their patriotic sensibilities and harmonized a powerful nationalist tradition with the prerogatives and putatively benevolent aims of the Cold War nation. According to this storyline, the pioneering efforts of the US space and Antarctic programs would engender on a global scale the blessings of freedom and prosperity Americans had long enjoyed owing to their history of frontier expansion.

Timely Currents of American Frontier Nationality

In the foreword to NASA's 1966 educational booklet *Space: The New Frontier*, President Lyndon B. Johnson made quick sense of the US space program. He urged students to think that "the characteristic American confidence in the future," which "brought the first colonists westward across the Atlantic to settle the eastern shores" and subsequent "generations westward across the continent to build up our country," would prompt their nation to set off for the space frontier.[1] Young readers were probably not surprised by his comments, for they lived in a society saturated with frontier references. If the backcountry had called on Americans' vigor and vision and paid them stupendous economic dividends, so too did scientific research, which the leading architect of federal science policy Vannevar Bush famously labeled "The Endless Frontier."[2] In the same spirit *Life* magazine described engineers as pioneers of the "Frontiers of Technology," while its sister periodical stretched the term as it headlined the "frontiers of jazz" and even the "frontiers of modern theology." *Time* magazine turned President John F. Kennedy's favored political slogan into shorthand for his government, which it simply called the "New Frontier."[3] Newspapers reminded readers of the world's dangerous frontiers separating allies from hostile enemies, just as America's western line of settlement, according to lore, had separated it from unruly Indians. Those war-whooping natives were then as common in American mass culture as the steely cowboys and cavalry who fought so hard to keep them on the reservation. Popular entertainment brimmed with stock frontier heroes and villains, casting them in ubiquitous fables of the Wild West and giving them occasional cameos in sagas of large-scale engineering and futuristic fantasies of space travel. While visiting the popular Disneyland, for instance, vacationers witnessed America born and raised in the amusement park's hardscrabble Frontierland and saw it mature among Tomorrowland's high-tech outposts across earth and in outer space. Spinning mythic tropes into sellable stories and commodities, mass marketers followed Walt Disney's lead and cast America as a pioneering nation forged on its western frontiers and destined to conquer the planet and move into outer space.

They disgorged their fare on tested ground, for space travel had long been a sellable theme with frontier overtones. Americans had often treated outer space in fanciful terms. All manner of exotic and sometimes hostile aliens, reminiscent of frontier beasts and bloodthirsty savages, appeared in popular entertainment. As early as 1845, a Philadelphia Minstrel Show featured a blackface character whose stereotypical slights were all the more comical for being committed before winged moon-maidens.[4] The amusement park impresario Frederic Thompson may have had minstrel and vaudeville precedents in mind when he came up with his wildly successful Coney Island ride "A Trip to the Moon." Unveiled first at the 1901 Pan-American Exposition in Buffalo, NY and imitated by traveling carnivals, Thompson's fantasy spaceship carried thousands to the moon, including President William McKinley, where they nibbled green cheese and enjoyed the staged antics of lunar midgets.[5] In the early twentieth century many imagined extraterrestrials were not so harmless and hokey. When the star struck savant Percival Lowell turned his astronomical observations into popular theories of defunct Martian civilizations, pulp novelists exploited that attention to hustle stories of extraplanetary adventure. Edgar Rice Burroughs's fame rested not only on his Tarzan novels but also on his bestselling "Barsoom" stories featuring John Carter as intrepid space trotter. After the gold prospecting Carter escaped an Indian ambush by mystical transportation to Mars, his fine balance of masculine hardiness and manly refinement, so characteristic of fabled frontiersmen, carried him through epic Martian battles. The medium that brought the similarly gallant Buck Rogers and Flash Gordon to life continued to pump out space fantasies into the 1950s, when Hollywood populated the cosmos with thinly drawn monsters whose taste for killing white men and abducting fair skinned damsels smacked of boilerplate Indians.[6]

A second stream of popular culture envisioned a space frontier not by filling it with pulp stand-ins for cowboys and Indians but by describing a realistic, sequential process of celestial exploration paralleling that of the New World and the American West. Among early harbingers of this tradition and the first technically reasoned account of space travel in the United States was Boston litterateur Edward Everett Hale's 1869 story about an artificial earth satellite serialized in *The Atlantic Monthly*.[7] Early science fiction "wonder stories," strongly influenced by Jules Verne's logical tales of space travel, enjoyed sizable audiences decades later. So too did the Urania Scientific Theater of New York's Carnegie Music Hall, which presented a Vernian "'scientific performance' entitled 'A Trip to the Moon'" in 1892. This wondrous new incandescent show featured a series of glass plate slides depicting a voyage beyond the earth.[8] Early moviemakers including Thomas Edison projected similar jaunts into space onto the silver screen, while popular magazines like Hugo Gernsback's *Amazing Stories*, known in the 1920s and 1930s for its sci-fi pulp, published thoughtful predictions of space-age science and technology.[9]

During the 1950s this tradition came of age as experts and media pundits soberly announced the United States would soon blast into the heavens. Inspired by atomic age technology and advised by leading aerospace engineers like the German turned US Army expert Wernher von Braun, whose V2 missiles went from terrorizing wartime London to touching the edge of space, they depicted in technical detail an impending age of interplanetary travel.[10] Just as lonely caravels and covered wagons opened New World frontiers, pioneering rocketmen would unlock the heavens for waves of space travelers.[11] Amateur rocket societies in the United States and Europe spent decades hammering out technical schematics, and their rough blueprints for a piloted expedition to the lunar surface were first rendered as a Hollywood blockbuster in the 1950 movie *Destination Moon*. Two years later the popular periodical *Collier's* followed up with seven cover-page stories on space exploration. In the series opener "Man Will Conquer Space Soon," the editor introduced "the story of the inevitability of man's conquest of space. What you will read here," he averred, "is not science fiction. It is serious fact."[12] That story did not feature rocket-riding cowboys or murderous aliens. It laid out as impending fact a staged conquest of space, a process one scholar has labeled the "von Braun paradigm."[13] The magazine suggested that a fleet of reusable winged rockets would service an orbital station that conducted scientific research and military reconnaissance and served as a port for nuclear-propelled ships bound for the moon and Mars. This paradigm found its biggest national audience through television. When Walt Disney produced a series of TV shows to promote his planned amusement park, he took a page from the *Collier's* series and based the Tomorrowland spots on the theme of space exploration.[14] Retaining von Braun and colleagues once again as expert advisors, he broadcast three Tomorrowland shows in the mid-1950s— "Man in Space," "Man and the Moon," and "Mars and Beyond"— that introduced nearly one hundred million viewers to what Disney called America's "new frontier, the frontier of interplanetary space."[15]

By the time the United States announced plans in July 1955 to launch a scientific satellite during the International Geophysical Year (1957–58), popular entertainment had prepared Americans to think of space as a new frontier, even if they did not expect the United States to pioneer that frontier anytime soon.[16] Public discourse about Antarctica was quite different. Although filmmakers and journalists saw US Navy and International Geophysical Year (IGY) expeditions as evidence of an impending international rush to occupy Antarctica, American audiences were not conditioned to think of it as a colonial frontier like outer space. Long before the IGY, Jules Verne spun a tale of south polar adventure, James Fennimore Cooper wrote a stirring novel about American sealers in Antarctica, and Edgar Allen Poe fictionalized natives there as exotic as any New World Indian.[17] But these stories and the published journals of Antarctic sealers, whalers, and research scientists did not romanticize the region as suitable for settlement. *Time* speculated as late as 1947 that in Antarctica "there may be a hidden valley heated to tropical balminess by volcanic energy [where] unknown fauna may be nibbling at unknown flora." But the magazine rightly noted that the continent "has been written off by most romancers as hopelessly, unromantically cold."[18] Testy political conditions there, upset by territorial competition and Moscow's newfound attention to the region, kept most US authorities in the 1950s from openly expressing their nation's interest in settling Antarctica. Without an official colonial policy, most Americans simply viewed the frozen continent as lacking the fertile fields to sustain national settlement or the primitive peoples in desperate need of America's "civilizing" influence.

Although they rarely envisioned colonies in Antarctica, Americans often treated it as a primeval frontier forever locked in an ice age but ripe in natural resources for the taking. The narrator of the 1948 movie *The Secret Land* spoke thus when he called it a terribly cold and lifeless land that nevertheless remained the "one untouched reservoir of raw materials left in the world."[19] Conditioned perhaps by their state's history of feverish speculative venture, Florida investors eyed this reservoir and formed the short-lived Antarctic Colony Associates in the early 1950s to prospect a gilded path in Antarctica. A *Christian Science Monitor* reporter likened them to earlier generations who struck out for the Klondike and predicted that Antarctica would be as rich as

Alaska "insofar as metals, minerals, and fuels are concerned."[20] This might have pleased the celebrated polar explorer Rear Admiral Richard E. Byrd, who foretold of a future bonanza of Antarctic resources in his secret deliberations with the National Security Counsel and in his many public addresses. While giving the keynote to a meeting of the Poultry and Egg National Board, for instance, Byrd told the country's most prominent chicken farmers that the United States needed the continent's "untouched reservoir of natural resources." He penned articles and took to the airwaves to excite other Americans as well about the treasure buried beneath Antarctica's ice sheets.[21]

Richard Byrd enjoyed credibility as a veteran polar explorer. In addition to blazing a trail to prospective resources, Antarctic explorers like Byrd assumed a national duty to conquer earth's remaining wilds and model for impressionable young men the manly virtues once exhibited by America's legendary backwoodsmen.[22] Richard Byrd seemed to have these traits in abundance and was showered with many national honors and ticker tape parades for his bold expeditions and aeronautical surveys of Antarctica in the late 1920s and early 1930s.[23] In a 1959 reprint of the Rear Admiral's memoir *Alone*, in which he described his grueling solo stint in an Antarctic outpost in 1934, the publisher praised him for pointing the way "toward an inner strength and fortitude the world needs now more than ever."[24] Such was the unflappable fortitude of Byrd. In the same vein, a chronicler of south polar exploration hopefully declared that even as "the number of unvisited places on the face of this planet has been reduced to almost zero" Antarctica remained an "infinite challenge to man's hardiness and courage."[25] Rear Admiral Byrd often spoke of Antarctica in these terms, as he did when he called his competition to choose a plucky Boy Scout to accompany the US Navy's 1956 Operation Deepfreeze to Antarctica "a real contribution in strengthening America spiritually, physically, and morally."[26] This had been a founding mission of the Boy Scouts of America since 1910, when it first encouraged young scouts to follow the example of renowned pioneers who fought so hard to tame America's wild frontiers. *National Geographic* magazine evoked this martial spirit when it called Deepfreeze "An All-Out Assault on Antarctica."[27] It was also the language favored by dozens of people who wrote to Byrd requesting a berth on that expedition, including one hopeful volunteer who expressed his readiness to endure "pain and hardship" so as to "do my part for mankind by helping to conquer new lands and civilize them."[28]

US officials rarely used the frontier analogy to describe their planned space and Antarctic projects in the lead up to the IGY. If not one-time affairs, these projects were not yet abiding national programs. After

IGY research stabilized Antarctic politics and Sputnik launched the space age, the United States rolled its south polar operations into the National Science Foundation's permanent USAP and its fledgling satellite venture into the brand new NASA. The creation of these lasting programs in 1958 was a turning point in American frontier nationality, when sober authorities began referring to Antarctica and especially outer space as modern counterparts to America's long settled frontiers. During NASA's decade-long barnstorming buildup to its triumphant lunar landings, the space frontier became a favored motif among politicians, aerospace professionals, and media observers. The motif often linked the cosmos to the legendary landscape of the American West. "We want to give the American people," NASA Administrator James Webb explained in 1962, "something in modern terms that they can be as proud of as the heroic march of the settlers who came West over the Oregon trail."[29] Webb's deputy Robert Seamans invoked a related trope when he compared the US space program more broadly to the discovery and settlement of the New World, the seminal act of America's pioneer experience. Seamans asserted that the "same drive that led Columbus to explore the outer reaches of the known world will induce modern man to explore the deeps of the solar system."[30] The space-age version of the nation's frontier mythology evoked not only its promise of geographic expansion but also its underlying telos, which Congressman George Miller (D, CA) neatly identified as Americans' relentless drive for progress. "What is most important is for America to experience a rate of progress in every aspect of its culture that is dynamic and questing," Miller argued, and he raved that its space program, "as no other national program has in the past, will affect every aspect of our society."[31]

Although they treated the space frontier as a commonsense motto and natural extension of America's pioneer history, their rhetorical embrace of the motif was not foreordained. Wanting to avert what historian Walter McDougall called a potential "orgy of state-directed technological showmanship that would be hard to stop, [and] might spill over into other policy areas," President Dwight D. Eisenhower avoided any discourse that warranted a budget-breaking human space program.[32] He preferred to speak about practical science and orbital operations and regarded Project Mercury, NASA's first man-in-space program started in 1959, not as preparation for interplanetary exploration but as a limited effort to study the fitness of men for orbital travel and military reconnaissance. Eisenhower's pick as NASA's first administrator T. Keith Glennan similarly emphasized "the early and direct benefits which we may anticipate from our investments in space technology" and disparaged "space cadets" whose scenarios of extraterrestrial travel could not "justify the expenditure of hundreds of millions that such ventures into space will cost."[33] The President's Science Advisory Committee (PSAC) admitted that with US prestige at stake it needed to take up "the challenge to transport man beyond frontiers he scarcely dared dream until now."[34] It nevertheless agreed with Glennan that NASA's piloted space program was hard to justify along scientific, military, or economic grounds. So too did President-Elect John F. Kennedy's transition team, which urged him to "diminish the significance of this program to its proper proportion" and try "to make people appreciate the cultural, public service and military importance of space activities other than space travel."[35]

Just weeks before President Kennedy shelved this advice and endorsed a monumental sprint to the moon, a *Reader's Digest* author criticized the still nascent Project Mercury as a "Senseless Race to Put Man in Space."[36] The quiet skeptics he surveyed soon went on record against the much pricier Apollo lunar landing program they believed diverted scarce funding from more practical space operations. Some eschewed the trope of the space frontier because they preferred military and commercial applications in earth orbit, while many scientists believed research satellites and planetary probes cost much less and promised far greater returns than piloted expeditions in space.[37] They often conceded that astronauts boosted national pride and prestige and were far more glamorous than robotic probes, but they insisted that starry-eyed fantasies of human exploration did not warrant the wasteful cost of encumbering spaceships with life support systems. Most outspoken was Philip Abelson, director of the Carnegie Institution's Geophysical Laboratory and editor of *Science* magazine, who condemned Apollo and regretted that "enthusiasts have described space as an enormous frontier of vast potential and...stated frequently that we face an opportunity similar to that of Christopher Columbus when he sailed to discover a New World." The "analogy is a poor one," Abelson told Congress in 1963, for practical payoffs would come from basic science rather than sending men to planetary bodies that were, romantic dreams aside, "less habitable than the most miserable spot on earth."[38]

President Eisenhower tried to check the influence of such spaceflight enthusiasts.[39] But they continued to push the frontier analogy, which gained traction in popular entertainment and among social commentators, government officials, and spokesmen for the aerospace industry. This was especially true after May 1961 when President Kennedy called for a speedy lunar landing before the end of the decade. "No single space project," Kennedy wisely predicted, "will be so difficult or expensive to accomplish."[40] Suddenly flush with money to chase the president's goal, a rapidly expanding NASA built several new centers and distributed contracts for university research and industrial development throughout the United States. According to one historian, the space agency "undertook a mobilization comparable, in relative scale, to that undertaken by the US to fight World War II." When its annual budget soared from roughly $500 million to $5 billion in the early 1960s, NASA's staff grew threefold to nearly thirty five thousand and its contractor workforce increased by a factor of ten, peaking in 1966 at more than four hundred thousand employees.[41] Its infrastructure stretched across the country and aided the economic modernization then occurring in the South and West. The millions who benefited provided ready ears for the visionary and profitable schemes of the space frontier. But its most prominent boosters were NASA officials and the myriad aerospace executives and professionals who steered this dramatic mobilization as well as scores of officeholders who hitched their political stars to the space program.

These supporters formed what political scientists call an "iron triangle" of state power and what historian Brian Balogh labeled a "proministrative" alliance of industry professionals, government administrators, and politicians.[42] This coalition was bound by their mutual interests and bankrolled, managed, and serviced the US space program. Just as these scholars theorized, this alliance cultivated the political and professional backing necessary to sustain a state bureaucracy devoted to space exploration. But its members also looked beyond Washington powerbrokers and made their case directly to the American people. By encouraging public support, what they often self-servingly called "public understanding" of their program, they hoped to secure steady financing for this costly federal effort.[43] As one ranking NASA official explained, the space program's "long term health and support as a publicly financed endeavor is dependent to a large degree on a significant increase in public understanding."[44] There were many reasonable justifications for that program and these proministrative actors publicly emphasized their varying priorities. Members of Congress reminded constituents about well-paying aerospace jobs in their districts. White House speechwriters celebrated the space program as a national investment in economic and military security, while State Department spokesmen commended it as an effective vehicle for international cooperation and development. Corporate executives pointed to their companies' profitable diversification into civilian space operations, and aerospace professionals focused on their many promising research projects. NASA officials made each of these points as they itemized the many ways their agency fulfilled its legislative mandate to make America a leader in outer space. These were the people physicist Freeman Dyson had in mind when he wrote in 1969 that "the ultimate strength of the space program derives from the fact that it unites in a constructive effort a crowd of people who are in it for quite diverse reasons."[45] That strength might have become a weakness had this coalition simply offered a confusing jumble of motives for the US space program. Citizens might then have renounced it as a pork barrel prize of narrow interest groups. Instead this constellation of program supporters found common and politically effective ground by folding their separate interests and wide ranging justifications into stirring tributes to exploration, national leadership, and human progress.

Wernher von Braun admitted as much. Such promotion "helped to arouse and sustain widespread support for a mammoth and expensive undertaking," he wrote in 1969, "which most likely would never have gotten off the ground on its scientific and technological merits alone."[46] Von Braun was a gifted promoter and like other advocates often framed his lofty claims about US space exploration in the familiar guise of the frontier. Owing to their expert authority and prominent public presence, NASA officials and their political and professional partisans turned a popular cultural trope into the apparently sensible motif of the space frontier. Unlike NASA, the National Science Foundation did not have charismatic advocates of von Braun's stature or so many interest groups ready to proselytize for Antarctic exploration. Nevertheless the NSF and its smaller band of government allies, professional partisans, and sympathetic journalists regularly spoke obliquely about a south polar frontier, describing the USAP in hackneyed terms as an inspiring exercise in manly courage and a pioneering effort to exploit the untapped riches of a wasted continent. They were occasionally more explicit, as the US Navy had been in the title of its 1956 report "Antarctica: The Last Frontier," and even depicted this last earthly holdout in the subsequent decade as a staging ground for the space frontier.[47] The frontier motif ennobled their parochial causes, but NASA and NSF officials and their supporters were not cynical schemers. Their favored motif resonated with their own patriotic sensibilities about America. Aligning space and Antarctic exploration with a nationalist narrative of frontier expansion, they assured themselves and others that the United States remained a special nation with a promising destiny despite its formidable Cold War challenges.

In so doing they trod a path most famously followed by John Kennedy. He used the metaphor of the frontier as complimentary shorthand for his presidential campaign and cast himself as an energetic visionary ready to rescue America from exhausted policies and waning national stature. When Kennedy accepted the Democratic Party's 1960 nomination he announced:

> I stand here tonight facing west on what was once the last frontier [where] the pioneers gave up their safety, their comfort and sometimes their lives to build our new West. They were not captives of their own doubts, nor the prisoners of their own price tags. They were determined to make the new world strong and free—an example to the world, to overcome its hazards and its hardships, to conquer the enemies that threatened from within and without.

Kennedy entreated Americans to follow their lead and face the "frontier of the 1960's" with similar grit, and he promised a "New Frontier" of government policies to help them revive the United States. The New Frontier was gauzy political parlance that recapitulated a heavily worked metaphor, but it persisted and followed Kennedy into office because it was a braided motif whose two primary tresses resonated deeply with the zeitgeist of the time.

One of those strands related to the tangible metrics of US economic power and international standing and the other pertained to the moral fabric of American society. Although Americans enjoyed unmatched wealth and global influence, many were distressed by recession and their fast changing economy and worried that a nuclear armed, spacefaring Soviet Union might convince potential allies that it would soon eclipse the United States. The vein of the frontier myth relevant here traces back to historian Frederick Jackson Turner, who nearly seventy years earlier named America's western frontiers the source of its vibrant economy and unyielding democracy. This was the frontier Kennedy called on when he promised "bold measures" to energize the economy and convince "a watching world" that a free and democratic United States "can compete with the single-minded advance of the Communist system." Turner's kinetic contemporary Theodore Roosevelt popularized the other main branch of the frontier myth. Concerned about social decline, he promoted the strenuous life of legendary frontiersmen as inoculation against modern men's enervating values and habits. Kennedy exuded Roosevelt's personal vitality and saw America's primary challenges in similar terms. Americans could reboot their economy and defeat the communist juggernaut, he advised, through principled national action, choosing "courage" over "complacency," and committing to "lead vigorously" rather than becoming a "tired nation." Weaving these two plaits together, Kennedy called on Americans "to be pioneers on that New Frontier" by pursuing uncharted opportunities and being "stout in spirit."[48]

The frontier motif had a long history, but it was not a timeless expression or a true reflection of national experience and character. It was so popular and applied so well to outer space and Antarctica during the late 1950s and 1960s because its two strands together then captured Americans' common hopes and allayed their widespread fears. As high-tech programs of scientific exploration, the US space and Antarctic programs evoked Frederick Jackson Turner's frontier by purportedly calling on individual creativity and the genius of democratic society to generate the knowledge, industrial innovation, and natural resources needed to boost the Free World economy. These programs also hailed the rejuvenating frontier of Theodore Roosevelt by offering an apparently drifting nation a noble purpose to fire up its citizens. These programs did so in real space, geographic frontiers which had existed since Turner and Roosevelt only in nostalgic fables of the Wild West. The new frontiers of space and Antarctica were hardly like the verdant New World and hard scrabble American West. But their vacant settings provided fitting opportunities for an Atomic Age nation to make even the most forbidding places serve humankind.

Democracy and Progress on Turner's Frontiers

Frederick Jackson Turner was a little known historian in 1893 when he delivered his groundbreaking paper "The Significance of the Frontier in American History." He soon became an influential scholar due to his thesis that America was molded by its successive frontiers. Turner's colleagues had offered well-placed Americans a favorable national story with which to identify by tracing their special dispensations to Teutonic ancestors. The United States was a democratic powerhouse, in their account, because select Americans enjoyed advantages inherited from their Anglo Saxon and Germanic parentage. Turner turned away from ethnically exclusive Old World roots to more inclusive New World frontiers as the source of America's unique history. As he succinctly put it, the "existence of an area of free land, its continuous recession, and the advance of American settlement westward, explain American development." And what a wondrous development that had been. Unlike thoroughly landed Europeans, Americans were on the whole free, democratic and prosperous due to their constant adaptation "to the changes involved in crossing a continent, in winning a wilderness, and in developing at each area of this progress out of the primitive economic and political conditions of the frontier into the complexity of city life."[49]

Turner believed the United States was a special nation, the pinnacle of civilization, because its endowment of unclaimed land helped it avoid

Europe's rigid social boundaries and drags on economic development. The evolutionary process of turning savage acreage into civilized heartland resulted in America's unprecedented individualism and democracy. Taming wild lands required an independent spirit, Turner wrote, "promoted equality among the Western settlers, and reacted as a check on the aristocratic influences of the East."[50] Typical of contemporary evolutionary thinkers, he drew selectively on British naturalist Charles Darwin's theory of natural selection and on the suppositions of the earlier French savant Jean Baptiste Lamarck. Whereas Darwin held that creatures best suited to their challenging natural environments were most likely to survive and pass on their traits to offspring, Lamarck suggested that living beings purposely adapted, by the grace of God, to the forbidding world around them. Turner shared Darwin's emphasis on competition when he declared that "at the frontier the environment is at first too strong for the man" and that "He must accept the conditions which it furnishes, or perish." But he leaned more heavily towards Lamarck's notions of hereditary adaptation and suggested that pioneering Americans developed and passed on to their successors the independence and democratic spirit they needed to survive on rough frontiers. Turner also embraced the Lamarckian idea of progressive change and felt that the frontier, by stripping "off the garments of civilization," suppressed its noxious aristocracy and pushed civilization forward by breeding the natural virtues of liberty and democracy. In short, free land made free men who built the world's most democratic, civilized society.

Turner also credited wild lands with stimulating Americans' "acuteness and inquisitiveness; that practical, inventive turn of mind" that made for their matchless industry. These favorable traits prepared pioneers for Darwinian competition with nature and helped them achieve a Lamarckian, progressive transformation of each frontier. Traders were the "pathfinder of civilization," in Turner's scheme, making primeval lands ready for pastoral ranchers and industrious farmers, whose improvements led to the formation of towns and "finally manufacturing organization with city and factory system."[51] This repeated process of turning putatively empty wastes into productive settlements kept the United States on an even keel as it matured into an economic powerhouse. America's hinterlands absorbed throngs of immigrants who would have depressed urban wages and working conditions, and they became home to families that outgrew heartland farms or exhausted their fertility. Thus "the sanative influences of the free spaces of the West were destined to ameliorate labor's condition," Turner theorized, "and to postpone the problem" of divisive inequality that beset other advanced societies.[52] The United States became a diversified economic power because new frontiers also furnished natural resources critical for industrial development, growing markets for manufactured goods, and productive outlets for surplus capital. In short, free land made prosperous men who built the world's most productive nation.

Turner made such a splash because he seemed to explain the nation's emerging problems. When he expounded his thesis shortly after the US Census Bureau announced the closure of America's last continental frontiers, the country was wracked by economic depression. This crisis struck many as foreign to Turner's expansionary nation, as were the agrarian populism, urban socialism, and often violent confrontations between industrial management and labor then prevalent in the United States. If his thesis was correct and these problems were caused by a lack of open land, then they would abate if the country found effective alternatives to its moribund geographic frontiers. Struggling workers would benefit again from what scholar Richard Slotkin calls the "populist" opportunities of Turner's frontiers. Whereas free land once did the trick, new frontiers would provide hard working people the chance to improve their lot and insure America's ongoing "diffusion of property, of the opportunity to 'rise in the world,' and of political power."[53] What's more, new frontiers would help assimilate the millions of southern and eastern Europeans then landing on America's shores. "In the crucible of the frontier," Turner wrote, "the immigrants were Americanized, liberated, and fused into a mixed race."[54] They would presumably continue to melt together only if the United States found substitutes for its then settled backcountry. These substitutes would also benefit the broader arc of the corporatized economy by reproducing what Slotkin labels the "progressive" aspect of Turner's thesis, which associated land conquest with "the steady transformation of small individual concerns into large economic and political institutions."[55] If America's turn-of-the-century problems resulted from stalled industrial growth, new frontiers would spur companies to expand once again, specialize further, and swell into more efficient and economy-boosting ventures.

In the ensuing years, homegrown imperialists urged America to maintain its inexorable progress by taking new territory. Many people who rejected such annexation still embraced a muscular foreign policy capable of securing global trading partners. If the United States had exhausted its continental frontiers, it would find in far flung corners of the world new resources, markets for its over-production of farm and industrial goods, and profitable outlets for its investment capital. According to historian John Mack Faragher, Turner "understood the connection that Americans made between the West and the world," even if he did not promote imperialism or economic empire, and he too sought an alternative to America's defunct western frontiers.[56] He found it in America's new public universities where enterprising men of talent could generate the science and technology needed to rescue the nation from its post-frontier doldrums. "General experience and rule-of-thumb information are inadequate for the solution of the problems of democracy which no longer owns the safety fund of an unlimited quantity of untouched resources." That being the case, Turner explained, the "test tube and the microscope are needed rather than the ax and rifle in this new ideal of conquest." Scientists and engineers who wielded these new tools "must be left free, as the pioneer was free, to explore new regions and to report what they find; for like the pioneers they have the ideal of investigation, they seek new horizons [and] are not tied to past knowledge."[57]

Many prominent figures exalted these modern frontiers of science and technology when the United States boomed once again during and after the Second World War. Vannevar Bush famously did so in his 1945 report *Science—The Endless Frontier*. In a letter of transmittal to President Harry Truman that evoked Turner's thesis, Bush wrote:

> The pioneer spirit is still vigorous within this nation. Science offers a largely unexplored hinterland for the pioneer who has the tools for his task. The rewards of such exploration both for the Nation and the individual are great. Scientific progress is one essential key to our security as a nation, to our better health, to more jobs, to a higher standard of living, and to our cultural progress.[58]

In Bush's estimation, research and development (R&D) required the individual creativity and democratic cooperation common among Turner's storied pioneers and promised the security and economic growth that had once sprung from these industrious homesteaders. This analogy was very common during the next quarter century. Nobel laureate and Chairman of the US AEC Glenn Seaborg used it when he drew parallels between modern R&D and the historic works of "men of great vision" who "believed in the frontier" and "saw clearly how science and engineering could develop the vast potential of the West."[59] So too did National Science Foundation chief Alan Waterman, who compared the "people who established this country and pushed its frontiers across 2000 miles of wilderness" with America's new pioneers who plied the "frontiers of the mind" so as "to meet the challenges of the technological age."[60]

The frontier analogy was popular in part because its denotation of progress offered hope that atomic age science and technology would not degrade or even destroy humankind. Americans should have rejoiced in an R&D complex that helped revive the domestic and foreign economies, defeat the Axis powers, and contain the Soviet Union. But many worried that complex had upended their traditional ways of life. Agricultural modernization had thrown millions of farm laborers out of work and automation raised the specter of chronic industrial unemployment. These economic woes paled alongside more existential anxieties as millions suffered psychological fallout from the nuclear arms race and worried with good reason that these weapons might suddenly destroy the world.[61] Glenn Seaborg remained optimistic that modern science and technology were "the most powerful forces for material advancement unleashed by man." He nevertheless gave voice to this anxiety and acknowledged "the very survival of modern civilization" hung in the balance.[62] President Eisenhower more graphically warned that nuclear war could result in "the annihilation of the irreplaceable heritage of mankind," condemning it "to begin all over again the age-old struggle upward from savagery." This was the struggle Americans replayed less apocalyptically on Turner's successive hinterlands, and proponents of Vannevar Bush's "Endless Frontier" hoped humanity could continue its upward ascent rather than fall back in a cascade of mushroom clouds. Eisenhower expressed this very wish in 1953 when he introduced his "Atoms for Peace" initiative, whereby nuclear material would be made available worldwide for civilian applications, and proposed that "this greatest of destructive forces can be developed into a great boon, for the benefit of all mankind."[63]

The prospect that science and technology were progressive frontiers offered therapeutic relief for people anxious about the atomic age. But this frontier motif was popular largely because it neatly fit the visionary outlines of the American Century; cutting-edge research and ensuing high-tech development bolstered Free World security and stimulated economic growth at home and abroad. Policymakers had dropped Depression-era plans for government management of an exhausted economy and used Washington's monetary leverage and fiscal outlays, including billions spent annually on R&D during the 1950s, to spur on a fast growing economy.[64] In Lloyd Berkner's 1964 treatise *The Scientific Age*, this prominent science advisor and academic research consortium chief called America's robust economy one of unprecedented "abundance." While less developed people "struggled for the bare necessities" and endured a "traditional economy of scarcity," Berkner avowed, Americans had nearly conquered poverty through new "adaptive industries" that thrived in their free and democratic society. These industries relied on "innovation derived from the most advanced science of our time" and promised "new sources for employment, wealth, and human satisfaction."[65] Like Turner, Berkner essentially treated science and technology as modern frontiers. The difference between countries saddled with traditional economies and the adaptive United States paralleled the Turnerian breach between its impoverished back-country and the rich heartlands they became. Americans had bridged that gulf by conquering a continent, and they now aimed to progress further and lift the less fortunate in their wake by pioneering the new frontiers of science and technology. The State Department said as much in 1961 when it reaffirmed US support for people worldwide who had "a new and urgent awareness that although the misery of man exists as a fact it need not continue to exist" since "scientific and technological gains give promise for them and their children of a better life."[66]

The United States was not the only country holding out that promise. So too did the Soviet Union, whose state publicists claimed that a spacefaring proletariat would soon vault past its capitalist foes with first-rate science and technology needed to create an international and egalitarian economy of abundance. "This is the meaning of Sputnik," Lloyd Berkner warned Congress in 1960, and he advised lawmakers to demonstrate the moral and material superiority of democratic capitalism by supporting leading programs of exploration in space and Antarctica. "In our day, when few physical frontiers remain, peoples visualize space and the Antarctic," he explained, "as challenges that must be accepted by a great nation to demonstrate its mettle."[67] His passing reference to "physical frontiers" was instructive, for these two realms gave Turner's abstract frontiers of science and technology a familiar, grounded expression. America's spacemen and Antarctic explorers conducted research and deployed new technologies not in prosaic laboratories but in the world's final geographic quarry. After decades in which America's only vacant lands existed virtually in popular westerns, it seemed the United States once again had spatial frontiers at its disposal. This was reassuring for a nation still shuddering from Sputnik, which had sharply undercut the conventional truism, as *Scientific American* publisher Gerard Piel put it, that "The free winds of liberty are vital to the life of science."[68]

The credibility of America's world leadership rested in part on this apparent truism. This is why Vice President Richard M. Nixon responded to Sputnik by insisting that free men still retained their "long-run" advantages even if "a dictator state can in the short-run achieve spectacular results by concentrating its full power in any given direction."[69] Since liberty and democracy entailed the sometimes sluggish give-and-take essential to science, Nixon counseled, Americans conducted better R&D even if they did not work as fast as their slave-driven enemies. His words were of little consolation due to the possibility that the Soviet

Union would surpass the United States in the short run and lord over it with a new weapon as threatening as nuclear bombs or Intercontinental Ballistic Missiles. Thus Nixon also exhorted Americans to strive as their forefathers had "from the earliest days of our history with the challenge of an unconquered wilderness and an apparently limitless frontier" so as to accelerate the R&D "as necessary to human progress as it is to the security of free men."[70] America's wild backcountry endowed pioneers with the individualism and democracy deemed essential to the scientific enterprise, but it also primed them to quickly conquer that frontier. Doing the same to the space and Antarctic frontiers would stimulate the cutting edge science and technology critical to its world leadership and demonstrate that a free and democratic nation could be as dogged and fast acting as the illiberal Soviet Union. Astronaut Edward White could have referred just as easily to Antarctica when he said US prestige was at stake in space because it entailed "a large scale program" which was "a test of our democratic system. If it is carried out as planned, it will prove that people working together under a free government can compete successfully with other systems."[71] NASA's T. Keith Glennan made a rare feint to a space frontier when he similarly urged Americans "to demonstrate once again that free men— when challenged— can rise to the heights and overcome the lead of those who build on the basis of subjugation."[72]

Glennan flirted with the analogy because the frontier myth signified the vitality of America's liberal democracy, but he generally avoided it due to his down-to-earth emphasis on practical returns of the space program. Such dollars and cents benefits, however, bolstered the motif of the space and Antarctic frontiers as well. Turner's "populist" frontiers gave legions of enterprising men opportunities to improve their lot. Although it was obvious that common folk would not be able to do so in short order in Antarctica and outer space, their long-term prospects seemed better. Observers projected Turner's frontier stages onto these forbidding realms; male explorers would be followed by pioneering families and then by towns and urban metropolises. Veteran Antarctic explorer Rear Admiral George Dufek noted in 1957 that although only intrepid men could then endure the dangerous continent, "women will come to the Antarctic" and settlements ensue as the United States built bases and airfields there, introduced atomic power, and even modified the polar climate.[73] Women were not yet there in 1963, but the United States Information Agency (USIA) was so confident that atomic power would facilitate their arrival that it rhetorically asked: "Will there one day be homes and schools and children in Antarctica?" America's operational base in

Antarctica's McMurdo Sound seemed poised for those homes and schools. According to the USIA, it was already a bustling outpost with a hospital, "friendly and conservative" neighborhood banker, "church with well attended services," and "everything any other town has except mothers, wives and sweethearts."[74] A visitor was so impressed by the size and amenities of the complex that he wondered, with no hint of irony, "How long will it be before the Winter Olympics are held at McMurdo Sound?"[75]

Antarctica often featured as an early step in the nation's parallel conquest of the space frontier. An NSF scientist declared that the "choicest test ground on the planet" for America's move "into a new environment outside earth's atmosphere" was "the land and water surrounding the southern pole."[76] Wernher von Braun thought so as well and determined that NASA could test equipment in Antarctica and model its lunar landing program on the USAP.[77] Commenting on the rocketeer's visit to his Antarctic station, a young researcher drew the parallel further and claimed there was "no need to wonder what a base on the moon might be like someday. We're it already."[78]

Von Braun had roughly sketched out such a base many years before as part of a broader plan for exploring outer space. The paradigm was faithfully followed in a 1958 Sunday supplement in national newspapers which explained how the United States would conquer the "New Frontiers" of space with an orbiting space station, a small lunar city, and piloted forays to Mars, Venus, and distant stars.[79] A stream of books, television shows, and movies appeared in these heady days of the early space age detailing similar voyages and a subsequent American charge into space. The TV series *Men Into Space* worked through each step of von Braun's scheme between 1958 and 1960 and finished confident that "as certainly as the sun will rise tomorrow, men—and women— will go hand in hand into space" and "colonize" other worlds.[80] CBS television's *The 21st Century* struck the same chord several years later. One episode declared that after "some lunar Lewis and Clark, men will walk on the moon" and "colonize" it, while another suggested that residents of "lunar colonies" would launch "huge space ships that may enable us to explore and then exploit each of the planets."[81] Stanley Kubrick vividly pictured this scenario in his 1968 blockbuster movie *2001: A Space Odyssey*, giving millions of people visual cues for what the federal US Information Service had tagged "the Future of Space Exploration" and President Nixon's Space Task Force subsequently proposed as a fitting post-Apollo program.[82]

This plot's frequent appearance in popular media and government reports as well as in reams of letters NASA received from would-be

astronauts attests to the seductive dream of the Turnerian space frontier. If this populist dream was at best a long-term prospect, there were
more immediate ways Americans purportedly benefited from their
nation's space program. A ranking advisor pointed to space-related
"wages, salaries, and profits to people throughout the country," earn-
ings for hundreds of thousands of people working directly for the program as well as countless more who serviced their many needs.[83] But
if American populism once emphasized broad opportunity for gainful
work, it had come during the twentieth century to connote something
very different. In a political economy oriented to corporate capitalism
and mass consumption, the path to the traditional populist ideal of
independence and dignity ran through substantial consumption rather
than worthy employment. "The activity of opening up new frontiers
has time and again stimulated economic development," a ranking
NASA official explained, and agency boosters emphasized the fact that
it was as consumers that all Americans, indeed all humankind, stood to
profit from the space frontier.[84] Skeptics wisecracked that the country
got only the orangey drink Tang and metallic lubricant Teflon from its
lunar expenditures. But the mantra of program supporters like agency
director James Webb was "What we do in space will have practical value
on earth" by generating wondrous new technologies that private enterprise alone could not have created. [85] NASA spinoffs included miniaturized circuitry that went into computers, avionics, home electronics, and
medical devices like hearing aids and pacemakers. Heat-resistant materials appeared in industrial furnaces and fabrics worn by firemen and
forest rangers. The program also brought forth compact power sources,
efficient insulation, stronger paints, and new alloys and plastics, as well
as the better known Tang and Teflon. Furthermore the meteorological and telecommunications satellites NASA helped develop during the
1960s touched people's lives the world over by revolutionizing weather
forecasting and global communications. [86]
     The practical benefits of Antarctica were similarly portentous if
more earthly than the high-tech spinoffs of the space program. Many
informed observers felt that in a resource-hungry world Antarctica
likely had minerals and nutriment that would, in President Johnson's
words, "serve the aspirations and the well-being of individual men in
all nations."[87] Thus *New York Times* science columnist Walter Sullivan
pointedly reported, "Antarctica is bound to have mineral resources
comparable to those of other great continents."[88] These resources were
especially vital given an NSF panel's 1957 Malthusian findings that
"national survival may depend" on the "discovery of new and hitherto
unsuspected sources of mineral wealth."[89] Worried more about the

rumbling stomachs of a burgeoning world population, natural history writer Roger Caras focused instead on the region's bountiful seafood when he advised Americans that "Antarctica is as much a part of your future and the future of your descendents as the farm crops not yet sown."[90] The NSF plainly agreed, asserting that with "rapid growth of population in all parts of the world, every source of food will be needed, and the abundant life of the southern seas may contribute notably to the future of mankind."[91] Antarctic authorities also forecast that south polar research would positively affect humankind by improving global communications and navigation as well as the prediction and even mitigation of severe weather events. Some anticipated that Antarctic icebergs would water the world's deserts and that the continent would host airports, food storage depots, and nuclear waste repositories.

The motif of the Antarctic and space frontiers further evoked what Richard Slotkin calls Frederick Jackson Turner's "progressive" emphasis on the growing scale and integration of American enterprise. Turner deemed America's evolutionary ascent from backcountry subsistence to industrial-age civilization a natural outcome of its frontier experience. Its forward momentum picked up again after the Great Depression and carried the country into a new stage of breakthrough development that Lloyd Berkner called the "Scientific Age." This progressive conceptual framework was in line with the contemporary suppositions of modernization theory. The many influential thinkers and policymakers who subscribed to that theory in the 1950s and 1960s held that Americans, who already enjoyed the world's most advanced consumer-oriented economy, would grow ever more prosperous as their society developed further. It would do so if they became more educated and scientifically literate and worked in ever more specialized, integrated, and competitive enterprises that depended on America's leading R&D complex. Echoing the Lamarckian rise from primitivism outlined by Turner, modernization theorists also believed that American aid, trade, and investment could save poorer nations from Soviet depredations and put them on a universal path of economic development and social progress pioneered by the United States.[92] The liberal internationalist project of the American Century, in historian Michael Latham's nice summation, thus entailed "defeating the forces of monolithic communism by accelerating the natural process through which 'traditional' societies would move toward the enlightened 'modernity' most clearly represented by America."[93]

America's frontier myth naturalized this particular historical process and framed it in a familiar nationalist vernacular. As applied to Antarctica and outer space, the myth implied that the nation's record of progress would automatically continue and rub off on the rest of the Free World. The southern continent once again had a lesser but still important role to play in this global evolutionary drama. The tangible metrics of a modernizing America, especially its impressive mobility and global reach, were evident in Antarctica. *Science* magazine brightly noted that "the sled dogs once imported to the Antarctic have been replaced by the motorized toboggan, the Snow-Cat, the helicopter, and the airplane," which was light years ahead of famed British explorer Robert Falcon Scott's ill-fated march to the South Pole a half-century before.[94] "Now American planes fly in along the route that Scott followed," the *New York Times* added, "taking the roughly 800 miles from the coast in one giant five-hour stride."[95] Responsible for these flights and USAP logistics, the Department of Defense (DOD) further sharpened the technology and techniques needed to bring far-flung environments into humankind's everyday orbit. Scientists tried to do so as well by integrating Antarctica into geophysical models that helped people understand and, many then expected, even control planetary forces. Those researchers not only added to the wealth of human knowledge, they educated American college students and enriched their universities' diverse professional specialties. The NSF made this possible by annually awarding millions of dollars in grants to Antarctic researchers, scholarships to their graduate students, and funding for their laboratories. All this was critical to the nation's wellbeing, director Alan Waterman explained, since its ongoing progress and standing in the world "depend to an increasing extent on the effectiveness of our research and development effort and on the number and quality of scientists and engineers which our educational system is providing." The USAP also helped America fulfill what Waterman called its "responsibility to help the developing nations to apply today's knowledge to the problems of underproduction, hunger, and disease," since Antarctic research and resources, once again, promised a bevy of benefits to the international community.[96] Thus the well-oiled relationship among government sponsors, industry contractors, and academic researchers in Antarctica exemplified the R&D complex that accounted for the nation's rapid modernization and its wealth and world power.

The USAP was a minor force for modernization compared to the space agency, whose high-tech spinoffs led to next generation products and industrial processes. US airlift in Antarctica demonstrated its rapid mobility and unrivaled global reach, but its thundering rockets most dramatically symbolized the nation's lightning advance. There were tangible signs of that rapid development across the United States, particularly around the NASA centers and aerospace enterprises that followed the postwar industrial migration to the South and West. These regions had been quiet agricultural hinterlands until World War II, when their new utilities, plentiful resources, and cheap land and labor made them favored locations for wartime industries. The president of the US Chamber of Commerce pointed out that America's aerospace complex continued this electrifying modernization of its southern "crescent of the sun" for the same reasons after the war.[97] By the late 1950s and 1960s, aerospace industries turned many agrarian communities "from Florida's palmetto thickets to the Texas ranchlands below Houston" into economically diversified boom towns, thereby shifting the nation's center of economic gravity toward these once sidelined regions.[98] While locals generally welcomed what *National Geographic* breathlessly called "a Space-Age Boom," modernization sometimes rubbed up against traditional cultures and customs.[99] Many offended locals cried foul, for example, when NASA's James Webb implied that Alabamans' provincialism and racial attitudes made it difficult to find professionals willing to work at the Marshall Space Flight Center in Huntsville.[100] NASA ultimately found its recruits among the expanded pool of specialists it helped create. Because the agency needed an army of already scarce scientists and engineers, it became a leading sponsor of university research and education. After it turned a pilot project into the Sustaining University Program in 1962, NASA helped modernize American universities by earmarking hundreds of millions of dollars for academic fellowships, research grants, and laboratory construction by the end of the decade.[101] It thereby played a far greater role than the NSF in building a gathering surplus of professionals needed to keep America's economy moving forward and its R&D driven industries ahead of foreign competitors.

Congressman George Miller cited these factors when he declared that the space program benefited all Americans, whether they were "the average housewife, or the Texas cattleman, or the Iowa wheat farmer, or the California fisherman." NASA enriched what Miller and modernization theorists called their "deeply integrated" society through product spinoffs, payrolls, and support for university research and industrial development as well as through innovative managerial techniques.[102] Like the wartime big science and technology project to build an atomic bomb, the space agency faced the novel administrative test of coordinating diverse personnel in a fast-paced technocratic undertaking across the United States. As one scholar points out, NASA operated in the public eye and its "diverse and diffuse objectives and its heterogeneous organization created more profound managerial challenges" than even the secret wartime Manhattan Project to build atomic bombs.[103] James Webb thought his agency's response was its most important contribution to the modernizing nation. In this unfolding technocratic era, characterized by what historian Walter McDougall calls "state-supported, perpetual technological revolution," new public-private collaborations that a Congressional staff study called "a mighty industrial and government complex" generated the cutting edge R&D capable of maintaining America's military and economic preponderance.[104] NASA's administrative innovations, the blended hierarchical control and decentralized authority of its so-called "systems management," helped managers of such large-scale endeavors adapt to changing political and technical conditions while coordinating a mighty phalanx of government employees, industrial contractors, and academic researchers.[105] "The essence of our job," Webb thus asserted, "has been that of organizing and managing the use of available knowledge and technology in a purposeful and effective way" so that "technocracy" served America's modern needs without overwhelming its free market and democratic traditions.[106]

If this conglomeration was a logical extension of Frederick Jackson Turner's progressive frontier, the physical imprint of that technocracy also smacked of his famous thesis. Turner believed not only that the frontier had molded America, but also that national progress was inscribed upon the landscape. The artist John Gast depicted this inscription well in his famous 1872 painting "American Progress" (Figure 2.1) As wild animals and Indians in the painting flee westward before waves of male itinerants and then family settlers, the female national symbol "Columbia" floats overhead and unfurls telegraph wire transmitting the pulse of urban civilization from whence she came. Had he painted in the 1930s, Gast might have depicted Turner's environmental makeover by showcasing the monumental dams that bridled wild rivers of the West. Ten years further and he would have likely painted that region's bustling technocratic complex that produced fissile material that atomic boosters like Glenn Seaborg believed could "compete with the forces of nature."[107] By the mid-1950s this competition was in full swing. So said *Time* magazine, which praised American scientists and engineers for taming unruly environments that had made man "a prisoner to his surroundings, starving in desert lands or drowned by torrential floods." The magazine hailed them for showing the world there was "almost no project too big to tackle, no reasonable limit to reshaping the earth to make it more productive."[108]

This was the environmental promise of the American Century. That promise was not a commitment to nature conservation and protection. Rather it was a vow to make every corner of the earth more

Figure 2.1   John Gast's 1872 painting "American Progress."

pacific and productive. Glenn Seaborg had this promise in mind when he called "the discipline of planetary engineering a practicable one" since Americans now had the "muscle to move mountains, possibly change the climate, and extract natural resources previously locked tight far below the ground."[109] John Gast might have then painted not simply a continental procession but a planetary project in which Turner's modern frontiers of science and technology moved mountains around the world. Gast might have done so because planetary engineering was not just Seaborg's technophilic fancy. It conformed to what historian James Patterson called a "guiding spirit of the age," namely Americans' "grand expectations" that with their vaunted science and technology they faced "no limits to progress."[110] This was the spirit evoked by TV newsman Walter Cronkite who, as host of CBS's *The 21st Century*, blithely predicted that Americans would soon have "the power to feed the world's billions," protect them from disease and inclement weather, and send them to live on other planets.[111] *Time* announced with similar aplomb they would soon "land on the moon, cure cancer and the common cold, lay out blight-proof, smog-free cities, enrich the underdeveloped world and, no doubt, write finis to poverty and war."[112]

News and entertainment media took their cue from trusted corporate and government authorities that also trafficked in these grand expectations. Summing up years of visionary boosterism, a 1957 Atomic Energy Commission (AEC) comic book about the atom pictured this "most constructive instrument of man's inventiveness" propelling ships and planes, revolutionizing medicine, agriculture, and industry, and modifying the weather. The atom's most fantastic application was powering cities sheltered under transparent domes "in the forbidding continent of Antarctica [and] the frontiers of space."[113] In subsequent years, the agency's "Project Plowshare" proposed using nuclear bombs as massive earth-moving devices and pushing the limits of planetary engineering even at these far reaches of the world.[114] Several corporate exhibitions at the 1964 New York World's Fair dallied with these limits as well and echoed the AEC's bold predictions for a rosy atomic future.[115] General Motors trumped them all with its Futurama II pavilion which showed "man conquering new worlds" by turning arid deserts into fertile fields and fetid jungles into industrial centers. Like the AEC comic book, the exhibit's most fulsome examples of planetary engineering were a large south polar station and bustling lunar base.[116] By carrying the seeds of civilization to what it called the "frontier lands" of Antarctica and the moon, these atomic-powered settlements illustrated the Fair's official hallmark and fitting catchphrase for planetary engineering: "Man's Achievements on a Shrinking Globe and in an Expanding Universe." [117]

Corporate marketers had peddled this sort of technological optimism at World's Fairs for several decades.[118] By the early 1960s, the federal government did so as well. When an expert panel determined that US science exhibits at the 1958 Brussels Exposition were too abstruse and failed "to awaken the U.S. public to the significance of the general scientific effort and the importance of supporting it," they recommended that subsequent displays "must appeal to the general public rather than the specialist." Accordingly the United States Science Exhibit at the 1962 Seattle World's Fair dramatized for public consumption how Americans then exerted "control of man's physical surroundings."[119] Two years later the United States Pavilion in New York explicitly linked these exertions with the nation's frontier past by anointing the scientists and engineers who were then mastering the icy poles and soaring into space as the progeny of Columbus and "the scout in the wilderness."[120] If these early pathfinders forged a great nation by taming the New World environment, then future pioneers would lift America and the Free World further by overcoming obstacles in the Antarctic and space frontiers. As President Johnson exclaimed that same year: "The challenge in Antarctica is similar to the challenge in outer space ... overcoming obstacles of nature which have frustrated and baffled the human race through the centuries."[121]

In a rational but also psychologically salutary prospect of turning earth-threatening atomic swords into plowshares, nuclear energy often featured as the force capable of overcoming these obstacles, even the most forbidding ones in outer space. "It seems almost predestined that the development of nuclear energy and our readiness to explore space have coincided," the AEC's John McCone thus intoned in 1959, since "in the field of missiles and space vehicles the atom is about to assume an indispensable role."[122] That prospective role began unfolding in 1949 when researchers first looked at atomic energy to power future reconnaissance satellites. The AEC and DOD subsequently developed a portable radioisotopic thermoelectric generator called Systems for Nuclear Auxiliary Power (SNAP). Unveiled by President Eisenhower in 1959 and displayed abroad as part of his Atoms for Peace initiative, SNAP powered nearly a dozen satellites during the next decade.[123] Although Americans then worried about fallout from nuclear weapons tests, reporters seemed unconcerned when a rocket carrying a SNAP devise blew up and scattered its radioactive material in 1964.[124] Journalists were similarly nonplussed about the fissile material on board the Apollo lunar landers, and they failed to report the threadbare measures the NASA, AEC, and Public Health Service put in place in Florida in case an explosion during liftoff showered radioactive particles over throngs of spectators.[125] Perhaps the SNAP and Apollo's nuclear casks were small enough to preclude public distress. More likely they evaded reportorial scrutiny because a society that still had grand expectations trusted experts to safely consummate what Glenn Seaborg hailed as "a 'marriage' that was bound to occur between Space and the Atom."[126]

In 1960 the PSAC recommended that marriage also include nuclear rockets.[127] The AEC and NASA had studied thermal nuclear propulsion and started work on Project Rover, later called NERVA, to develop the technology. Echoing a decade of popular speculation about interplanetary nuclear rockets, the *New York Herald Tribune* reported in 1959 that the AEC-NASA project meant that ships to Mars "may be propelled by atomic explosions."[128] An aerospace engineer who understood that such rockets would be propelled by atom-heated gases rather than atomic explosions confidently predicted that "it is the energy of the controlled thermonuclear reaction which will provide us with the power resources to make possible a large-scale migration into space."[129] A ranking NASA official's upbeat expectation that the United States would send nuclear rockets to atomic powered bases by the twenty-first century appeared in public forums and agency planning documents throughout the 1960s.[130] This visionary prospect was not adequately far reaching for hydrogen bomb designer Edward Teller, who lived up to his reputation as an atomic booster by dreaming of a cosmic Project Plowshare in which the United States blasted out caverns for spacious lunar settlements with nuclear bombs.[131] Glenn Seaborg called this very scenario "Planetary Engineering: Phase II."[132]

What the atom could do in space, it could also do in Antarctica. After visiting the region in 1956, Walter Sullivan reported that "the atomic age is forcing us to reappraise our attitude towards Antarctica" since atomic icebreakers would soon glide through its pack ice, nuclear reactors sprout up there, and isotopes melt through the icecap and expose underlying rock to researchers and prospectors.[133] These predictions reflected the thinking of his naval escorts who had already concluded that the "construction and operation in the Antarctic of a nuclear power plant is feasible."[134] In early 1962 the Navy installed the first of three planned 1500 kilowatt reactors, dubbed "Nukey Poo," to provide electricity, heat and desalinated water at its Antarctic operations center and announced that "the atomic age reached McMurdo Sound."[135] Veteran commander George Dufek lauded it from retirement as a "revolutionary step in polar exploration" and the beginning of "a dramatic new era in man's conquest of the remotest continent."[136] For ten years Nukey Poo was the highlight of tours for visiting businessmen, government officials, and journalists. Although one such reporter waxed over this "feat of engineering," National Science Foundation chief Alan Waterman was more equivocal.[137] He granted that the reactor enhanced "the national prestige of the United States," but Waterman quietly worried it would hurt Antarctic science by taxing logistical support and possibly disrupting local research by elevating background radiation in the area.[138] His suspicions proved closer to the mark than puffed up public acclaim. Despite AEC claims that its portable reactors could be easily erected anywhere "in less than 90 days," Nukey Poo's containment vessel required major excavation and the agency tested the "turnkey" reactor for two years before it finally went on line.[139] Its team of 25 highly trained tenders reduced the Navy's operational flexibility, leaving military officials secretly ruing over "difficulties encountered in assembling, testing and achieving reliable operation with that plant." Within a year they decided to scrap the other two planned nuclear reactors for Antarctica.[140] Even after the Navy discovered in 1966 that Nukey Poo had leaked irradiated coolant, Americans heard only good news about what the USIA had celebrated as a blessed confederation of "the atomic age and the ice age."[141] When the Navy finally shut the reactor down in

1972 and carried 12,000 tons of tainted Antarctic fill back to the US, it asserted that America had demonstrated its "ability to operate a nuclear power plant safely and reliably in a remote, hostile environment."[142]

If nuclear power was a critical tool for planetary engineers, weather modification was one of their primary goals and a widely anticipated sign of their gathering power to make the earth do their bidding. When Congress directed the NSF to study weather control in 1958, US researchers hoped they could soon fine-tune rainfall, suppress crop-destroying hail, disperse traffic-halting fog, and defuse coastal hurricanes and inland tornadoes.[143] Cloud-seeding outfits then operated without proven success around the world, and many scientists remained skeptical about weather modification. But the National Academy of Sciences (NAS) and NSF endorsed it in 1966 as a promising field of research.[144] The federal Interdepartmental Committee for Atmospheric Sciences then deemed the "financial and other benefits to human welfare of being able to modify weather" so great it urged Washington to significantly expand funding for weather control research.[145] Owing to these expert endorsements, CBS's 1968 TV show "Can We Control the Weather?" answered yes, it "seems certain that man, at long last will soon do something about the weather."[146]

By virtue of the advanced R&D occurring in these new and final frontiers, outer space and Antarctica predictably featured prominently in this aspect of planetary engineering. The Navy's 1963 cloud-seeding trials over Antarctica did not bear fruit, but proponents of weather modification hoped the United States would eventually tame the hostile polar climate.[147] The 1968 documentary *Antarctica: Coldest Continent* gave voice to these hopes and heralded "an era of unlimited power when science may be able to change the temperature balance and convert the cold regions to productive areas."[148] Although Antarctic specialists were rarely so confident about finagling the region's weather with atomic power, NSF staff scientist Henri Bader thought that researchers could practice manipulating climates by using a nuclear generator to heat a thermal oasis on the remote polar icecap.[149] While Bader regarded the area as a weather engineering test site, many observers thought that basic climate research in Antarctica would lead to weather modification elsewhere on a grander scale. Since the continent exerted a substantial influence on global climate, one author noted in 1965, weather control around the world depended "to a surprising degree" on meteorological research there to fill in "a big gap in our jigsaw-puzzle picture of the earth."[150] Among Futurama II's scale-model displays of "the enthralling world of tomorrow," scientists working in "the International Weather Communications Center of Antarctica" were filling in that very gap.[151]

Visionary weather control schemes usually relied on orbiting devices. Some borrowed the idea from Hermann Oberth, the grandfather of German rocketry, of using huge mirrors circling the earth to manipulate weather on the ground and even warm up the polar regions.[152] These mirrors would constitute the "most important single development to evolve in the next hundred years from space exploration," Congressman Lionel Van Deerlin (D, CA) opined in 1963, by enabling "to a very great local extent at least, the control of the weather."[153] NASA and the DOD seriously considered the idea of orbiting mirrors several years later, albeit with the less benign intent of taking away the night time cover of darkness from the enemy in Vietnam.[154] NASA officials ultimately bet that meteorological and communications satellites, rather than mirrors, would help Americans achieve "improved forecasting, and perhaps at some future date, control of the weather."[155] Shortly after "Anna" became the first hurricane to form under the watchful eye of a US weather satellite in 1961, federal science advisors recommended a "World Weather Watch" system, a global network of ground stations and satellites to improve forecasting and lay the "scientific basis for exploring the possibilities of a large-scale modification to weather and climate."[156] President Kennedy had just endorsed such a system during his September 1961 address to the United Nations when he proposed "further cooperative efforts between all nations in weather prediction and eventually weather control."[157] Fresh from his historic July 1969 landing on the moon, astronaut Neil Armstrong showed that grand expectations for planetary engineering remained high at the end of the decade as well. While speaking with President Nixon through the window of a quarantine booth, Armstrong commiserated over a rained out football game and quipped: "We haven't learned to control the weather yet, but that is something we can look forward to."[158]

As the first men to set foot on an extraterrestrial body during the Apollo 11 mission, Neil Armstrong and Edwin "Buzz" Aldrin became modern icons of the mythic American pioneer. When *Time* magazine suggested that their lunar "steps may soon become a path, and the path a highway," it echoed Frederick Jackson Turner's thesis that such pioneers made way for gathering waves of frontier settlement.[159] Their historic mission also evoked Turner by demonstrating the vitality of an atomic age democracy in such rapid progress that it could deliver new worlds and radically improve the earth, including its weather, for the benefit of all mankind. But as Turner described the sweeping pageant of frontier development, he rarely focused on individuals such as Armstrong, Aldrin, and their partner in lunar orbit Michael Collins. The storied pioneers to whom they were often compared were the gallant frontiersmen exalted instead by Theodore Roosevelt, whose backcountry heroes invigorated their race and lifted the nation through their steely nerves and daring deeds. These were the protagonists of popular stories about the Wild West, and they were the heroic types who drew cheering crowds when the Apollo 11 astronauts toured the United States and scores of foreign countries. Thus supporters of US space and Antarctic exploration used a frontier analogy to promote their favored programs that summoned Roosevelt's popular mythology as much as Turner's thesis. In fact the analogy could not have otherwise endured, for the Rooseveltian heroes of the space and Antarctic frontiers held out hope that Americans could retain their cherished manhood and individualism even as they built a modernizing technocracy, in a space age version of Turner's frontier thesis, capable of securing and enriching the Free World.

Roosevelt's Rejuvenating Frontiers

When Frederick Jackson Turner introduced his frontier thesis, Theodore Roosevelt was in the middle of writing *The Winning of the West*, his four-volume work on America's westward expansion. The two scholars were close enough in mind that Roosevelt commended Turner for putting "into definite shape a good deal of thought that has been floating around rather loosely."[160] Turner returned the favor, according to historian Christopher Lasch, by being "generous in his praise and sparing of criticism" of Roosevelt's books.[161] Each complimented the other for expressing a shared belief that the United States was a world power and the most civilized society owing to its expansion into western lands. Their scholarly kinship was otherwise limited. Turner nostalgically treated the closed frontier as a lost engine of socioeconomic progress, while Roosevelt regarded it as ongoing model for societal rejuvenation. The democratic nation was propelled forward, in Turner's account, by nameless legions of industrious settlers who turned America's vacant backcountry into productive heartland. These laboring husbandmen took a back seat in Roosevelt's telling to rugged individualists, backwoodsmen like Daniel Boone and Davy Crockett who slew fearsome animals and imposed frontier justice on terribly cruel Indians. The West was not merely empty land put into commercial service by hard-working plebians. It was hostile territory heroically wrested, in Roosevelt's words, from "the most formidable savage foes ever encountered by colonists of European stock."[162] These colonists and their Indian-fighting descendents finished off three centuries of what he called "race expansion" associated with "the spread of English-speaking peoples over the world's waste spaces."[163]

Roosevelt assumed that Americans of Teutonic descent had inherited racial advantages from these expanding Old World peoples. But he was also a proponent of New World dispensation and believed those advantages became manifest when these racially superior Americans fought invigorating frontier wars against "savage" Indians. Roosevelt celebrated these pioneering individuals as agents of race progress and of American civilization. Although he did not use the term, Roosevelt was a Social Darwinist who loosely applied Charles Darwin's evolutionary theory of natural selection to human society. Roosevelt thought people competed with one another not only for nature's limited bounty but also for social power. His sociological take on Darwinism held that the fittest among them would come through this cutthroat social competition, earn privileged positions in society, and pass on their advantageous traits and well-deserved wealth and status to off-spring. Social Darwinists were also Lamarckian, for they believed these successful people would not merely survive; they and their fortunate progeny would rise to the top of society, in this case American society, and help it advance. Roosevelt and his intellectual peers assumed these successful people were primarily men of superior races rather than random individuals. Because their male ancestors had passed on ever more advanced traits acquired during successful social competition with other races, they inherited the capacities to carry modern civilization forward. These men of primarily Anglo-Saxon and German ancestry purportedly enjoyed the self-control and intellectual talent for political philosophy, economics, and science and engineering needed to keep civilization on the march. These characteristics distinguished them from more passionate and less brainy races, as did the fine-spun qualities of their highly evolved but differently constituted women. Unlike their male counterparts, these Teutonic flowers of female virtue would have wilted under the duress of social competition and public affairs. But they were esteemed, morally elevated partners in the racial project of civilization. They reproduced the race and graced its children and men with the moral refinement befitting their elevated station.

Their success was never finally secure, however, for their very gentility could be a competitive disadvantage against more muscular and fecund races. Roosevelt and fellow eugenicists worried that well-bred women would contribute to "race suicide" if they forsook the duty of ample reproduction so as to shower more luxuries on fewer children.[164] He felt the fate of civilization precariously rested on gentlemen as well, and he feared that if they too were beguiled by materialism, intellectual development, or manly control of their passions they could suddenly be overrun by less civilized yet more hard-driving races. Popular histories of the day indicated that this sorry fate had doomed ancient civilizations centered in Athens and Rome, which had grown fat and complacent before being overrun by rougher, more aggressive peoples. That fate had been avoided in the New World because America's better men, born with the pluck and talent to tackle hostile frontiers, were hardened and vivified in the process. As they recapitulated their brutish evolutionary origins by fighting Indians on the savages' terms, they fortified their manly inheritance and cultivated refinement with primal vigor and forestalled a life of ignoble ease by answering a higher calling, the martial pursuit of frontier conquest.

Theodore Roosevelt helped popularize rather than invent this frontier mythology. Its central figure remained the most celebrated hero of American popular culture during much of the twentieth century. He appeared rough drawn many years before in colonial narratives of Indian assault and abduction and was well developed by the early nineteenth century in the persona of Natty Bumppo, the protagonist of novelist James Fenimore Cooper's "Leatherstocking" tales (Figure 2.2). According to literary scholar Richard Slotkin, Bumppo's

Figure 2.2   A young Theodore Roosevelt projecting frontier vitality in 1885.

innate nobility and unflappable toughness as backcountry hunter and just executioner of vicious natives enabled him "to make the wilderness safe for a civilization in which he is unsuited (and disinclined) to participate." Cooper's backwoods genre took off and his hero became an iconic type reproduced innumerable times by dime novels, popular artists and writers, and William "Buffalo Bill" Cody's wildly successful Wild West Show. Buffalo Bill starred in his gun-blazing reenactments of frontier conquest and imitated Bumppo by calling himself a hero "to whose sagacity, skill, energy, and courage … the settlers of the West owe so much for the reclamation of the prairie from the savage Indian and wild animals, who for so long opposed the march of civilization."[165] Roosevelt's frontiersmen were thus already conventional types, and the future president catapulted to national office after casting himself as a similarly daring envoy of civilization. He achieved celebrity for dabbling as a western ranch hand and sheriff's deputy, exerting his cowboy bully in various government offices, and commanding men of stock frontier types during the 1898 Spanish-American War in his militia of "Rough Riders," a name borrowed from the Wild West Show. So renowned was Roosevelt's battle against Spanish colonial barbarism in Cuba that Cody's Wild West Show returned the honor by reenacting the Rough Riders' lion-hearted charge up Cuba's San Juan Hill.

A self-promoting model of virile frontier manhood, Roosevelt urged Americans to follow his lead and keep their vital heritage alive. Whereas Turner saw symptoms everywhere of a postfrontier socioeconomic crisis, Roosevelt saw signs of a torpid civilization whose natural leaders had forsaken their pioneering paternity. He urged these Anglo-Saxon men and white ethnics who had proven their racial mettle to protect civilized society from internal threats and foreign barbarism by pursuing what he called "the strenuous life" of muscular vigor and high purpose exemplified by backwoodsmen. In the absence of real hinterlands, his peers could toughen up and exercise away the nervous disorders medical professionals then thought debilitated over-civilized men through substitutes for frontier conquest. Roosevelt famously encouraged Americans to do so by hiking, mountain climbing, and big game hunting in the protected western parks and forests he championed as president. Boys and milksoppy men could hone their reedy muscles and weakened nerves closer to home through rough but well-regulated contact sports. "In a perfectly peaceful and commercial civilization such as ours there is always a danger of laying too little stress upon the more virile virtues," Roosevelt wrote, and he believed the resulting tough bodies and steely nerves protected the nation from the complacency, conformity, and materialism common in highly developed societies.[166]

The tranquility and wealth those societies afforded were hard won and would endure only if men remained fit and answered the call of national service. Americans could no longer do so through invigorating frontier wars, Roosevelt admitted, but they could still serve "the cause of civilization" with a "brave and high-spirited" foreign policy involving colonial administration in the Caribbean and Pacific, digging the Panama Canal, and building up the nation's navy and commercial empire. Roosevelt warned that if they stood idly by and "shrink from the hard contests where men must win at hazard of their lives and at the risk of all they hold dear, then the bolder and stronger peoples will pass us by, and will win for themselves the domination of the world."[167]

A half century later, Americans feared they faced a new contest for world domination with the bold but not yet stronger communist peoples. The United States retained a strategic advantage from its superior science and technology and Americans still appeared to be, in historian David Potter's words, a "People of Plenty" whose unique liberalism and material wellbeing was built on their abundant resources and individual resourcefulness. Since that now unrivaled abundance was widely evident in what economist John Kenneth Galbraith famously called with some critical irony an "Affluent Society," many Americans presumed that their cherished liberty and prosperity would rapidly spread if other nations followed their lead.[168] Belying this confidence and self-satisfaction, historian Warren Susman explained, brewed an "age of anxiety" about the personal costs of America's prosperity, something that had worried Theodore Roosevelt many decades earlier.[169] That anxiety accounted for the popularity of sociologist David Riesman's 1950 book *The Lonely Crowd*, which contended that Americans increasingly tailored themselves to social expectations and norms. Unlike nineteenth century "inner-directed" Americans endowed with strong characters and unwavering moral compasses, Riesman's "other-directed" peers relied less on "internal piloting" when interacting with others than on a "rapid if sometimes superficial intimacy with and response to everyone."[170] Simplified in the national press as an indictment of American conformity, Riesman's more subtle distinctions clearly struck a nerve. So too did the social strivers described in William Whyte's 1956 bestseller *The Organization Man*, who similarly lacked the self-direction of the hardy souls who built the nation. Instead of embracing their "Protestant Ethic" of independent labor, thrift, and competitive struggle, organization men exhibited a new "Social Ethic" of conformity to corporate and bourgeois norms. Social critic Vance Packard carried this line further in *The Hidden Persuaders* (1957), in which he argued that Americans' sacrosanct individualism was in question due to growing and manipulative power of advertisers and mass marketers. The venerable historian Daniel Boorstin had little doubt that Americans' autonomous selfhood was in fact waning and the "art of self-deception" waxing as they accepted "a thicket of unreality" sowed by public relations men and politicos and by the journalists who repeated their illusory hype.[171]

As long as the United States maintained its military and economic predominance, Americans kept these simmering anxieties about social conformity, organizational life, and enervating materialism at bay. However their confidence was rudely shaken in October 1957 when the Soviet Union launched Sputnik. The communist satellite seemed to indicate that a profligate people of plenty had fiddled amidst their affluence while America's security and prestige burned before enemy rockets. The nation's punditocracy echoed Theodore Roosevelt when it spoke of the potentially dire consequences of this staggering blow. The prominent theologian Reinhold Niebhur saw in Sputnik an "old historic situation," one that put a Rooseveltian spin on his famous warnings about a dangerous and depraved Soviet totalitarianism. Just as intemperate empires of antiquity were overrun by more muscular and brutish peoples, Niebuhr wrote, Americans now faced communist "barbarians, hardy and disciplined, ready to defeat a civilization in which the very achievements of its technology have made for soft and indulgent living."[172] Analogous language appeared in a congressional report warning Americans in 1959 that failure to answer Moscow's challenge would, "as the history of the complacent, wealthy, and unresponsive nations of the past attests, very probably point to a new dark age."[173] The salience of Roosevelt's frontier mythology was still evident several years later when Representative George Miller reminded his countrymen that other "prosperous nations, which dominated the world of their age, fell before the virility and single-sightedness that so strongly motivated the barbarian hordes."[174]

Many people also took Rooseveltian swipes at frivolous materialism to explain why the United States failed to be first in outer space. An outraged rocket engineer accused his peers of being "a smug, arrogant people who just sat dumb, fat and happy, underestimating Russia."[175] His was not a rare outburst. A gruff Harry Truman emerged from presidential retirement to blame Americans for being "fat and lazy and wanting too many cars and too many fancy gadgets."[176] Even US Chamber of Commerce president Erwin Canham, who normally gushed over the wealth generated by free market capitalism, acknowledged that "the haze and miasma of materialism" had swept over his country. Soviet leaders should not be "misled by the emphasis we appear to place on leisure—on the development of labor-saving devices and material comforts," Canham nevertheless added, for the United States "has by no means been drained of its virility" and is not "an old and self-satisfied society which is ripe for collapse."[177] Accustomed to darker biblical prophecies, the reverend Billy Graham was not so cocksure. He warned President Eisenhower that Americans "are growing soft" and their "amusements and greed for money is acting as a sedative." Preaching the need to "toughen up!," Graham urged "compulsory scientific training in all our schools, as well as compulsory physical training for all our young people."[178]

As Graham called for a physical revival—to complement his signature call for a religious revival— a broad consensus formed around another aspect of Roosevelt's "strenuous life," the need for Americans to give themselves over to a higher national calling. Before President Kennedy popularized the sentiment and promised what one historian called "the return to masculine hardiness" with his stirring inaugural words, "ask not what your country can do for you—ask what you can do for your country," this was the bottom line of his predecessor's Commission on National Goals.[179] Although that Commission hailed freedom and individualism as America's fundamental virtues, it summarily concluded in 1960 that the "American citizen in the years ahead ought to devote a larger portion of his time and energy directly to the solution of the nation's problems." More pointedly, "Americans must demonstrate in every aspect of their lives the fallacy of a purely selfish attitude—the materialistic ethic."[180] Henry Luce made this argument years earlier when he asked his countrymen to bear the noble burden of the American Century, and he repeated it at the dawn of the Space Age. "Sputnik should remind us," Luce wrote, "that any great human accomplishment demands consecration of will and a concentration of effort."[181] Alan Waterman picked up this thread and warned that "the U.S.S.R. shows a determination and a national spirit on the part of the people which seems to be relatively absent from the American scene." Chocking this up to a rich society that had complacently neglected its competitive fiber, Waterman called on Americans to identify national endeavors that roused the "latent vigor and enterprise our forefathers showed."[182] With those forefathers in mind, Theodore Roosevelt had selected martial pursuits reminiscent of frontier conquest. So too did Lloyd Berkner, who favored exploration of the last "physical frontiers" of outer space and Antarctica as bracing projects "that must be accepted by a great nation to demonstrate its mettle." The Washington establishment was primarily fixed on the space race after Sputnik, but it determined that America's reputation as a hale and hearty nation depended on its efforts in Antarctica as well. Presidential science advisor James Killian was so encouraged by these strenuous national efforts that he specifically cited the US space and Antarctic programs in 1960 as "evidence that the forces of innovation and aspiration, always strong in our society, even in periods of national fatigue and relaxation, may be ready for release."[183]

If *The Winning of the West* had once been the lofty aim that released those vital forces and bucked up the nation, Americans efforts to conquer space and Antarctica reassured Killian that his countrymen had found a fitting calling to muster their dormant verve. His colleagues on the PSAC believed that calling was to follow in outer space a vitalizing human impulse, "the compelling urge of man to explore and to discover."[184] This cardinal ideal of exploration, the indwelling urge for discovery and human progress, became a standard justification for a program whose primary purpose was to enhance national security and prestige. By concealing those prosaic interests and naturalizing what was in fact a nationalist impulse behind the program, this ideal gave flight to the common and related metaphors that space exploration was a "great evolutionary step for man," that it evinced the "same drive that led Columbus to explore the outer reaches of the known world," and that it was akin to the conquest of America's western frontiers.[185] Hence a *Hartford Times* columnist argued that Americans should go to the moon since their "sacrifice for an ideal," the "conquest of a new frontier," would do for them what the West did for earlier frontiersmen: "The challenge [of conquering that frontier] operated on their souls and made them aspire to the limits of their strength."[186]

The same "urge to adventure which will carry men to the planets," a NASA official avowed in 1961, "carried men on a traverse across the Antarctic Continent via the South Pole."[187] It was their chosen mission of scientific exploration in what Kennedy tellingly dubbed "the last great physical frontier of our planet" that prompted President Johnson to applaud "the dedicated and hardy citizens who strive selflessly for new knowledge in the cold and darkness of the Antarctic night."[188] Their naval escorts were just as praiseworthy, Rear Admiral David Tyree attested, because they gallantly won "a battle against bitter cold, against blinding storms, against pounding and frozen seas...through sheer determination, skill, and doggedness." When several sailors were accidentally struck down taking on Antarctica's unforgiving environment, the service channelled the spirit of Roosevelt by eulogizing each for having "served and died for his country just as devotedly and with as high purpose as if he had died fighting to preserve freedom."[189]

The motif of the space and Antarctic frontiers drew strength from the perception that valiant astronauts and polar explorers, like battle-hardened backwoodsmen, simultaneously served their nation and the higher cause of human freedom. At first glance the motif's martial implications threatened the analogy since these new frontiers were supposedly arenas of peace and universal progress. Once again, despite national leaders' adamant profession to the contrary, the United States operated in outer space and Antarctica first and foremost for national security and prestige. Its Antarctic program enhanced US naval capabilities and made it the foremost power in that geopolitically fraught region. Satellites provided essential reconnaissance and supported armed forces command and control, and astronauts paved the way for piloted military missions. Even the PSAC frankly acknowledged the vital "defense objective" of the US space program.[190] This apparent contradiction between a peace-loving and sword-wielding country actually worked to the frontier motif's advantage. That dissonance had long been smoothed over by Theodore Roosevelt's take on America's pioneering nationality, which sanctioned America's military action on its frontiers as a righteous defense of civilization. According to this nationalist narrative, battles for continental expansion were the Manifest Destiny of a peaceful nation roused to defend civilization by the surprise attacks of barbarous enemies. This storyline applied well to US participation in World War II and to its race into space. The unexpected orbit of Sputnik required a martial response as honorable as those following the Japanese ambush of Pearl Harbor and Indian sneak attacks on frontier outposts. In the logic of this frontier mythology, the United States was not an aggressive warmonger, as Kremlin propaganda insisted, but a reluctant warrior roused to protect the Free World with military defenses in space. The countenance of unarmed military men piloting spaceships and operating icebreakers and aircraft in Antarctica gave added force to the frontier motif by evoking Roosevelt's feint that America's frontier warriors strove for peace while their savage foes thirsted for war.

The fact that Americans who struck out for space and Antarctica were part of enormous state bureaucracies might have been another strike against the frontier analogy. After all such technocratic programs cut against the grain of fabled frontier individualism. The scale of such efforts and their "complexities of government, of finance and of organization" certainly distressed a *Life* magazine columnist in 1958 who wondered if Americans' "adventurous spirit" could endure when "Few frontiers can still be tackled by lonely individuals."[191] Even the frontier of science, which popular periodicals still inhabited with researchers "who resembled the traditional American frontiersman," began to look much less individualistic and adventurous.[192] Physicist Merle Tuve of the Carnegie Institution for Science put it bluntly in 1959 when he impugned America's R&D complex for treating scientists as other-directed organization men and turning them into "herds of giant research robots."[193] But the frontier motif gained strength from its psychically soothing implication that any hard-driving man of talent could still have a brush with adventure for a noble cause in space and Antarctica.

Thus in 1960 NBC television hailed the US Navy men working on the southern continent as common citizens who were uncommonly hardy and brave. These "grizzled veterans and fresh recruits, surgeons and soldiers" were ordinary Americans, having "come from across the 48 states [and] down country roads and from city bars."[194] Agency handlers and media outlets correspondingly treated astronauts as both typically wholesome Americans and rare heroes endowed with what novelist Tom Wolfe called "the right stuff" of steely nerves and steady hands. NASA tried to conceal their rougher edges by grooming them for press conferences and tamping down their profanity during live-broadcast space missions. Journalists were complicit in this public make over and did not write about the astronauts' bawdy off-duty antics. Instead *Life* magazine used its exclusive access, which it paid for handsomely, to help create the squeaky-clean public persona of the Mercury astronauts. "In spite of their extraordinary qualifications the Astronauts have many of the preoccupations, and even the small weaknesses, of more ordinary men," *Life* sentimentally noted, including "the condition of the grass in their yards and proper schooling for their children." Lawn mowing and childcare appeared once again in the magazine's special coverage of the Apollo 11 astronauts. They might have transcended the humdrum experience of their countrymen, but these lunar explorers came off as down-to-earth family men, morally grounded by quaffed wives and well-behaved children.[195]

Americans could relate to the astronauts, like the popular comic book figure Clark Kent, as workaday people like themselves. Of course the astronauts were not run-of-the-mill organization men. Like Kent's alter ego Superman, they appeared to be a cut above and conveyed people's Rooseveltian fantasies of personal heroism and daring deeds. Thus a spirit of adventure palpably filled the air during much of the 1960s as Americans vicariously thrilled in the hair-raising exploits of these pioneering men of steel. "We may have missed the opening of the West and the first trip along the Oregon Trail," a Boeing Aerospace brochure accordingly explained, "but we were a part of the greatest adventure of all time. We were there when man went to the moon," for which "there was a renewal of the spirit and a lift in national prestige as a result."[196] The romance of the astronauts emanated from the courage and rugged individualism that made Charles Lindbergh a national hero decades earlier. Just as Lindbergh single-handedly flew across the Atlantic Ocean in 1927, signaling that death-defying men could still be self-directed pioneers in the aeronautical era, astronauts evoked a manly independence for the technocratic Space Age.[197] The oft-made comparison between astronauts and Lindbergh was apparent when the reclusive aviator famously visited the crew of Apollo 8 and was on hand when Apollo 11 lifted off for its brush with history. *Time* ruefully acknowledged the limits of this comparison and admitted in 1969 that the "astronauts often seem to be interchangeable parts of a vast mechanism." The magazine nevertheless insisted that these brave men, like Lindbergh and Roosevelt's mythic pioneers before him, were in fact "essentially loners" reliant on their own pluck and talent.[198] Although their confining capsules were largely beyond their control, astronauts enjoyed what one historian deemed a reputation as self-directed "helmsmen" who combined "the pioneering image of '150 years ago' with a forward-looking mastery of technological change."[199] A college president who drew this very "analogy between the problems of the pioneers of the Western Frontier and the pioneers of outer space" insisted in 1962 that "both ventures require personal fortitude of a high order, integrity, courage and perseverance." He added that "the elaborate equipment of a Mercury Capsule is nothing without a John Glenn."[200] Astronauts' reputation as rugged individualists was evident a decade later when President Nixon honored the Apollo 13 crewmen after their aborted, near fatal lunar voyage. Nixon thanked the three men, who had turned their damaged capsule into a four-day cosmic lifeboat, for "remind[ing] us in these days when we have this magnificent technocracy, that men do count, the individual does count."[201]

Owing to the glamor of these spacefaring heroes, NASA was inundated with fan mail and international attention. The agency opened several of its centers to torrents of tourists, received innumerable requests to join the corps, and sent astronauts on the domestic lecture circuit and goodwill tours around the world.[202] Presidents who wanted to share their glory awarded astronauts medals in televised ceremonies, while lesser politicians begged the space agency to dispatch astronauts to accompany them to constituent events. One jilted Congressman testily advised NASA officials to remember his legislative vote on the agency's annual budget when fielding his next request.[203] Among the many volunteer associations that venerated the astronauts, the Veterans of Foreign Wars minted its own space medal and the Boy Scouts of America issued a "Space Exploration" merit badge in the early 1960s. Worried about delinquency and academic slippage, the Washington area Boy Scouts also started a local "Explorer Space Program" in the hopes that childhood preparation for the astronaut corps would inspire teenagers to stay on the straight and narrow and excel in school.[204] NASA's public affairs office took seriously the charge of educating and inspiring children. When it drastically cut astronauts' time consuming public appearances in 1962, the office made sure they still put in "appearances at youth group meetings."[205] Astronaut John Glenn for one believed it was his duty to cultivate that excitement, particularly among American boys, and he said that he worked with groups like the Boy Scouts to "encourag[e] young people to set goals and objectives and to take a more active part in such organizations."

The motif of the space and Antarctic frontiers gathered strength from that boyish excitement. The inspiring model set by hardy astronauts and Antarctic explorers appealed to many Americans who then worried, as Theodore Roosevelt had, that young men were growing soft in body and spirit. Moved by a perceived crisis in American manhood, a nun complimented John Glenn for displaying "the manly traits" of courage, self-control, and principled national service. She felt Glenn thus exemplified the very "'American Image' we wish them to portray in their adult lives."[206] The NSF's Alan Waterman similarly admired Glenn and his fellow astronauts to whom he compared the nation's Antarctic scientists for having the same manly poise and "restless desire to find new worlds to conquer."[207] Americans commonly treated those new worlds as Roosevelt had depicted America's western frontiers, too dangerous for women who constitutionally lacked the requisite strength, steady nerves, and urge for adventure. De facto prohibition against female participation in the US space and Antarctic programs reinforced this chauvinism, for women had no opportunity to prove if they actually had the "right stuff." Women worked for the space agency as secretaries, nurses, and number crunchers, but they found the astronaut corps off limits to them.[208] NASA public affairs director Julian Scheer explained in 1963 that there were no female astronauts simply because "no woman has yet met all the stringent qualifications which NASA has established for astronaut trainees."[209] However since these criteria included experience as a jet test pilot, a job reserved for experienced male military aviators, even the most qualified women could not become astronauts.

When medical screeners at the Lovelace Foundation determined in 1959 that women could in fact handle the rigors of space travel, the national press laid bare the gender chauvinism at the heart of Roosevelt's frontier mythology by mocking the idea that petticoats could lead the charge into space. The *Philadelphia Inquirer* begrudged that a "young unmarried woman...may make a fine test pilot for a simulated space ship. But wait until the simulation is gone," it joked, "and with it all chance of seeing the boyfriend for an indefinite time, and watch the old pioneer spirit evaporate into space."[210] When Lovelace investigators subsequently put scores of women through physical and psychological tests it developed for the space agency, 13 passed what was essentially the astronaut selection process. Several of these so-called Mercury 13 women stirred up enough fuss that the chief of NASA's manned space-flight program and two astronauts traveled to Capitol Hill in 1962 to explain that the astronaut corps would remain a male domain for the foreseeable future since NASA already had trained a team of trained astronauts and costly hardware designed for their male anatomy.[211]

When Russia's Valentina Tereshkova became the first female to travel in space the very next year, it looked like American women might soon break into that domain after all. Moscow had scored a major propaganda victory, according to feminist-minded women like Clare Booth Luce, who implored Americans to "stop trying to make paper dolls of our women" and respond in kind by training female astronauts.[212] After a brief public debate, the space agency remained unmoved. "No one denies that women will one day go into space," NASA's Julian Scheer explained, but the space agency had an ample roster of astronauts, and commonsense still held that women needed to wait for hardier men to test the safety of outer space and open that frontier.[213] People who accused Tereshkova of exhibiting feminine weakness by being "much too excitable" in orbit believed her flight confirmed conventional wisdom that women could effectively take to space only after men tamed that frontier.[214] That wisdom had been on display in the short-lived TV series *Men Into Space*. In a 1959 episode of that show, an American space commander determined that "before any colonization of the moon can be considered, we must investigate, under controlled conditions, the reactions of a woman to the moon-space environment." Although his test subject was initially much too excitable, even hysterical due to the rigors of space, she regained her domestic, feminine equilibrium by donning makeup and long dresses, cleaning the moon base, and cooking home-style meals. Satisfied that her lunar homemaking made the "first experiment of sending a woman to the moon a complete success," the commander triumphantly declared that "as certainly as the sun will rise tomorrow, men— and women— will go hand and hand into space."[215]

The National Science Foundation took a different tack and wanted to bring female scientists to the southern continent as early as 1959, but the US Navy refused to make that happen for another decade.[216] Officers professed the impossibility of building separate living quarters for women on ships and in Antarctica since operations there were still so new and the region so forbidding. A ranking navy man indicated in 1964 there might be other reasons for that reticence when he privately expressed opposition "to having women in Antarctica because of the kinds of problems which could arise should any incidents develop."[217] Rear Admiral Lloyd Abbot later admitted that one of these problems would have been a drop in discipline and morale among enlisted men competing for women's affections. An NSF official further revealed that the armed service worried that women could become pregnant, a real concern in so remote a location, or that they would soften the prospect of manly heroism that attracted enlistees to Antarctica.[218] The Navy was not alone in treating Antarctica as a rare challenge for American manhood. Jennie Darlington did so as well. Darlington accompanied her husband to Antarctica in 1947 and became one of the first women to visit and winter over there. She later confessed that "exploration is a male compulsion, generally beyond feminine comprehension."[219] Alan Waterman took this gender-specific compulsion on faith and challenged "those who believe that the age of adventure is dead or that science is for sissies [to] ask themselves whether they would be willing to brave the Antarctic weather at the South Pole."[220] Apparently not all scientists bravely answered the call, for a USAP official demeaned feather bedded men who were unwilling to work in Antarctica. "We have never been able to get a first-rate psychiatrist, psychologist, or sociologist to winter-over in Antarctica," he regretted, "not one of them has been man enough to leave the comforts of his wife's bed."[221]

Finally, while Roosevelt's nationalist mythology fit neatly with the prefeminist conventions of the late 1950s and 1960s, its racial prejudice potentially threatened the viability of the frontier motif. Impelled by the Cold War, government officials designed the US space and Antarctic programs to enhance national prestige not only by demonstrating America's superior science and technology but also by projecting its tireless efforts to enhance the wellbeing of all people, regardless of nationality or race. US State Department personnel advertised these programs to convince people in Africa, Asia, and Latin America that America was a color-blind leader of the Free World. Thus its global leadership entailed the basic presumption that the United States was a civic nation rather than an ethnic polity, constitutionally based on lofty principles and universal laws rather than narrow race and ethnic privilege. It was this civic vision of the human race, rather than Roosevelt's preferred Anglo Saxon race, to which an essayist referred when he insisted: "Dominion over the physical world belongs to the race and it is our common humanity that should reap the benefits of extending our sway beyond this narrow globe."[222] Presidential speeches and popular culture alike evinced this spirit of civic nationalism and depicted America's pioneering conquest of the space and Antarctic frontiers as efforts by all mankind and for the whole "Family of Man."[223]

This civic ideal may have been the public face of American national identity and Cold War internationalism, but it grated against the still deeply entrenched racism at the heart of American society. After all, legalized discrimination reigned in the American South until the mid-1960s. Despite new federal civil rights and immigration laws that forbade such discrimination, racial prejudice and de facto discrimination continued thereafter throughout the country. Demeaning stereotypes of black people popularized a century earlier still appeared occasionally on television and in movies, as did unflattering depictions of bloodthirsty Indians kept in check by movie star cowboys and cavalry like John Wayne.

These stereotypes began to flag as the United States experienced a traumatic period of conflict over civil rights, but their persistence exposed a still powerful strain of ethnic nationalism and racial chauvinism. Although American scientists and navy personnel operating in Antarctica were overwhelmingly light-skinned men and the astronaut corps was exclusively white, they were still admired by people at home and abroad as envoys of humankind. This was especially true for the astronauts, who enjoyed great acclaim in the dozens of countries they visited. They were thus a new kind of frontier hero who accommodated the civic ideals of the American Century but also embodied the shaky remnants of traditional racial nationalism. African-American women worked behind the scenes for NASA crunching numbers, and *Ebony* magazine featured black men trained in emergency medicine who supported the astronaut corps.[224] But the celebrated astronauts remained white and likely offered Americans who clung to racialist attitudes about their nation a comforting reference to Roosevelt's increasingly anachronistic hero, his Teutonic frontiersmen.

Conclusion

In the heady days following the first lunar landing in July 1969, NASA's head of manned spaceflight George Mueller declared "there remains for mankind the task of deciding the next step" and asked, "Will we press forward to explore other planets, or will we deny the opportunities of the future?" During the late 1950s and 1960s, the answer to Mueller's question was unmistakable—the United States should press forward and conquer the space frontier. The answer was obvious because proponents of US space exploration elevated a trope of popular culture into a conventional and credible motif, which they used to market their favored program and speak to the Cold War hopes and fears of Americans. They created a feedback loop in which mass media aped their expert predictions, while they in turn repeated what appeared to be commonsense discourse about the space frontier. By decade's end, this particular feedback loop collapsed as the trope of the space and Antarctic frontiers lost its cultural resonance and instrumental value. Trying to rescue this animating vision, Mueller warned that if Americans forsook the pioneering "spirit of our forefathers then will man fall back from his destiny, the mighty surge of his achievement will be lost, and the confines of this planet will destroy him."[225]

Mueller clung to a recently popular frontier motif that had effectively cast America's Cold War commitment to outpacing the Soviet Union in space and Antarctica not as a rash and costly break with national tradition but as a methodical plan in keeping with a pioneering past. The vein of the frontier myth associated with historian Frederick Jackson Turner suggested that America's liberal values remained strengths rather than liabilities in this confrontational age fueled by weapons of mass destruction and fast-paced R&D. Shades of Turner's thesis also popped up in claims that US space and Antarctic exploration would enrich humankind, push economic development around the world, and inscribe that progress upon the globe as a modernized United States pursued planetary engineering. Theodore Roosevelt's frontier mythology was equally evident, and its emphasis on martial defense of civilization resonated among security-conscious Americans and foreign allies alike. Thus a Brazilian newspaper heralded the 1961 suborbital flight of astronaut "Gus" Grissom as a "heartening comfort for the destiny of civilization and the predominance of democracy over totalitarian powers."[226] The frontier trope drew further strength at home from President Roosevelt's vision of societal rejuvenation. As Sputnik aggravated Americans' perennial modern anxieties about social conformity, materialism, and waning manhood, astronauts and Antarctic explorers provided contemporary models of the bully president's heroic frontiersmen. These hardy men of lofty purpose and bootstrap success embodied the nation's civic values at the same time they projected a white face associated with its tradition of racial nationalism.

The Rooseveltian strand of American frontier nationality collapsed in the late 1960s under multiple strains, most obviously the mounting pressure at home for racial civil rights. If the frontier motif's racist tilt became a liability, so too did its gender chauvinism. When NASA began training women astronauts in the late 1970s, the space agency did not treat them not as petticoat domesticators of the space frontier. It presented them and their African American male counterparts as brave and brainy. Furthermore, if "the social function of Apollo was to sustain a pre-Vietnam dream of conquest," as one historian contends, that war's tragic denouement thrust another dagger into the heart of Roosevelt's martial frontier mythology. That dream also drew strength from Turner's thesis that America, and now its Free World allies, progressed by conquering vacant lands. But as modern environmentalism erupted around the world, this Turnerian impulse to improve wild nature was discredited by a gathering mass movement to protect the natural world. Finally, the bright hope that a nation forged on its continental frontiers would domesticate the most forbidding corners of the planet and move into outer space darkened as America's once formidable economy sputtered and its global trade position weakened. The buoyant optimism of an economic hegemon gave way to national melancholy and soul-searching, which were ill-suited to the brash spirit of both strands of frontier nationality.

What had been a meaningful motif and commonsense way of describing US space and Antarctic exploration suddenly came off to many as naïve fantasy. Over the next two decades proponents looked for new and effective ways to market these federal programs. By virtue of their structural differences and varying profiles in public imagination, advocates and observers treated those programs very differently over the next two decades. At first, public officials and space program supporters largely dropped the frontier motif. Owing to that motif's deep cultural resonance and revived political utility, however, they revived the trope of the space frontier in the final decade of the Cold War. The paradigm of American nationality, expressed in discourse about the USAP, changed dramatically during this time. This program came to symbolize the nation's benevolent world leadership as steward of an endangered global environment rather than as pioneering explorer of distant frontiers.

Chapter 3

Antarctica and the Greening of America

After years of social unrest and political crises at home and of waning international power, many Americans welcomed the 1976 bicentennial anniversary of the United States as an occasion for rosy nostalgia. They staged readings of the Declaration of Independence and reenacted Revolutionary War battles, cruised in vintage sailing ships, and boarded a "Freedom Train" that toured the country. Like the first Freedom Train in the late 1940s, the bicentennial version exalted a history of homegrown liberty and displayed such treasured Americana as George Washington's copy of the Constitution. Although the rolling exhibit also carried a moon rock retrieved by Apollo astronauts, it dwelt largely on the past and ventured little into the troubling present or a futuristic space age. Patriotic nostalgia and cautious soothsaying were also evident that year at the opening of the Smithsonian Institution's National Air and Space Museum. President Gerald R. Ford dedicated the museum by invoking a national past in which "the frontier shaped and molded our society and our people," but he avoided airy talk of new frontiers as he vaguely foretold a future in which Americans "are always at the edge of the unknown."[1] The president struck a similarly prosaic note in his bicentennial salutations to the US Science and Technology Exposition at NASA's Cape Canaveral in Florida. Noting that its science, technology, and space program constituted "another chapter in America's history of reaching out to the unknown that began when the first colonists set sail across an unfriendly sea to an unexplored continent," Ford simply avowed that the United States would continue to make a "unique contribution to the progress of mankind."[2]

Little more than a decade earlier, the 1962 Seattle World's Fair depicted America's salutary contribution in the far more optimistic manner common at the time. One exhibit promised that the "highest aspirations of 20th-century man" would be realized as Americans designed a "radiant realm of Tomorrow" replete with "jetports, rapid-transit monorails and highways over which electrically-controlled automobiles ride on air," as well as factory farms robotically tended under protective domes. The General Motors pavilion at the 1964 New York World's Fair was similarly certain that Americans would erect glass and steel domes far afield and make the world's remaining wastes, including the "frontiers" of Antarctica and outer space, productively serve humankind.[3] President Ford avoided such exuberant futurism in 1976 because the buoyant faith in America's power evoked by the Fairs, and grand expectations that its science and technology could positively remake the world, markedly declined over the intervening years. The decline was so dramatic that White House staffers nearly overlooked the prospects for American science and technology when planning their bicentennial events. They hastily arranged the Exposition at Cape Canaveral to fill this "one major gap in Bicentennial preparations," but rejected as dated what they called "the traditional Buck Rogers glass-and-steel vision," on display in Seattle and New York. They proposed instead a "special vision of the future" based on "a human-scale back-to-nature vision—technology amidst trees and wildlife, American technology facilitating a return to nature."[4]

This Arcadian vision of technology working in harmony with nature was strikingly different from the technocratic dream of planetary engineering so recently in vogue in the United States. The motif of the space and Antarctic frontiers depended on that dream and the associated belief that Atomic Age America could positively transform the southern continent and move boldly into outer space. But the Bicentennial Exposition showed that this mindset was out of touch with what one political scientist then called a shifting "climate of opinion" in the United States. The "mounting public concern over an 'ecological crisis'" he identified accounts for the Expo's back-to-nature vision of science and technology.[5] The Expo's retreat from the conventions of planetary engineering and frontier conquest signaled an even broader change in public opinion, a mounting pessimism about the future that economist Robert Heilbroner called Americans' weakening "sense of assurance and control."[6]

Owing to tectonic shifts in domestic affairs and world events during the 1960s, Americans' collective assurance in the impeccable virtue of the United States and in its unassailable power had indeed waned. So too did their grand expectations that it could briskly impose its will on the natural world. Many who had exalted science and technology as boundless frontiers now questioned whether R&D could preserve their nation's flagging world leadership, mitigate mounting environmental problems, and sustain burgeoning populations. Vice President Nelson Rockefeller decried this pessimism in a bicentennial pep talk. He urged his pioneering countrymen to demonstrate once again "how a virile, adventurous people of many backgrounds and many views can achieve a consensus for liberty, freedom, and progress."[7] But a university president offered a darker measure of public opinion that year when he regretfully spoke of "the last gasp of the frontier thesis" and the end of America's long "quest for unlimited expansion."[8]

As public invocations of Americans' pioneering nationality waned in the late 1960s and 1970s, so too did spirited talk about the space and Antarctic frontiers. National officials and program managers had used the frontier motif to promote US space and Antarctic exploration and boost national pride and prestige. They and the media observers who repeated that motif responded to the country's changing mores and international position by speaking with more restraint about the mundane benefits of these programs. This practical storyline reflected Americans' diminished expectations and the downsized policies of the United States, which one historian aptly called a "disoriented giant."[9] As that giant regained its bearings in the 1980s, however, prominent arbiters of the public sphere described US space and Antarctic exploration once again as evidence of America's can-do spirit and millennial ambitions for human progress. While NASA promoters and chroniclers did so by resuscitating the trope of the space frontier, advocates of the US Antarctic Program (USAP) settled on a very different narrative of national character and purpose. In response to novel challenges to America's interests in the region and new priorities for research there, they primarily depicted the United States as steward of the south polar environment rather than conqueror of an Antarctic frontier. This revised national mission was succinctly captured in a 1979 documentary on Antarctica, which concluded: "Once our challenge was to conquer this final frontier, today our challenge is to preserve it."[10]

Twilight of the American Century?

At the end of the International Geophysical Year (1957–58) (IGY), the United States turned its temporary research projects in earth orbit and Antarctica into permanent programs that enhanced America's security and bolstered its battered reputation after the Soviet Union launched the Sputnik satellites. Since these high-tech scientific programs were newsworthy and vital to US prestige, NASA and the National Science Foundation (NSF) managed them in ways that reflected favorably on the United States. Public officials and program supervisors described them as pioneering efforts to conquer the world's final frontiers. Media observers similarly suggested that these costly programs carried on the nation's frontier experience and revealed its benevolent aim to bring about a better age, an American Century graced with a growing community of peaceful and prosperous nations. This was a relatively easy sell that resonated with Americans' fulsome embrace of their nation's pioneering character and frontier heritage. That lofty discourse also expressed their Cold War idealism about the benign application of America's awesome power. It offered an appealing alternative to many gritty realities of global leadership, namely America's massive militarization and frequent support for corrupt and authoritarian regimes. Seemingly untainted by such realpolitik, the country's push to make the vacant frontiers of outer space and Antarctica serve humankind seemingly manifest the expansive ideals of the American Century.

That sell became much harder in the late 1960s when Americans reevaluated their ideological commitments and discovered the limits of US power. Many had fervently believed that global communism emanating from Moscow posed an existential threat to the United States and Free World aspirants. According to this Cold War orthodoxy, the USSR was a totalitarian empire taking advantage of political volatility around the planet to impose its warped ideology and impoverishing governance on a world yearning for liberty and prosperity. Soviet hegemony in Eastern Europe combined with the revolutionary ardor of Red China, North Korean aggression, and spread of Third World socialism seemed to confirm this wisdom and galvanized public support for America's costly policy of containing communism. The Soviet Union stockpiled nuclear weapons and tried to match America's military might and global influence, but its stand down during the 1962 Cuban Missile Crisis signaled Kremlin recognition of America's superior nuclear forces. While Eastern Europe lay behind an Iron Curtain, American arms, aid, and trade helped keep Western Europe beyond Moscow's sphere of influence. Communists took control of China in 1949 and exacted heavy American losses on the battlefields of Korea in the early 1950s. But the United States checked China's influence in East Asia by fortifying South Korea and helping Japan become a powerful industrial democracy. Resistance to US internationalism also appeared throughout the so-called Third World in the late 1950s, which counted many new countries that yearned for economic autonomy and welcomed Soviet assistance. Nevertheless the United States traded with many of these new nations and it enjoyed productive partnerships with them in the United Nations. When economic nationalism and breaches in global containment seemingly imperiled its interests, the United States often handily neutralized those threats. Washington used covert support of military coups to abort nationalist reforms of Iran's oil industry in 1953 and of Guatemala's large land holdings in 1954, and it lent weapons, military advisors, and ultimately hundreds of thousands of US troops to stop communists from controlling all of Vietnam.

After two decades of hard-won successes, America's containment policy frayed as its military and economic preponderance abated. The backbone of that policy had been its strategic nuclear forces. These budget-straining weapons helped deter Soviet attacks against the United States and its global allies and prevent Moscow's putative client in China from similar action in Asia. But the Soviet Union achieved parity with those nuclear stockpiles in the late 1960s. Americans could no longer afford or psychically endure their effort to maintain nuclear superiority, which incidentally had not prevented taxing proxy wars from breaking out across the Third World. Thus President Richard Nixon pursued détente and arms control talks aimed toward nuclear parity with the Soviet Union. Nixon's even more stunning outreach to China in the early 1970s, which had angrily split with its erstwhile Soviet ally, signaled a new and more pragmatic Cold War posture. The United States appeared ready to treat communist countries as rational powers with divergent national interests rather than parts of a wild-eyed global monolith. America's influence in the Third World also waned substantially at this time. Members of the Non-Aligned Movement of less developed nations had questioned United States foreign policy in the UN General Assembly since 1955, but majority support for the United States in that body frayed over the next decade as many new and radicalized nations rebuffed Washington's international agenda. One of the most outspoken critics was revolutionary Cuba, whose pro-Soviet government led by Fidel Castro antagonized the United States, nationalized American holdings on that island nation, and then survived US-supported counterrevolutionary action in 1961.

If Cuba was an embarrassing breach in America's containment line and ended its streak of secret regime-changing operations, the grinding war in Vietnam dramatically exposed the limits of US power. American officials had worried that a unified Vietnam led by Ho Chi Minh, an anti-colonial leader who was also a communist, would upset US regional interests by nationalizing its economy or joining a growing bloc of communist countries. Thus after embattled French rule finally ended in 1954 and the region temporarily split in two, Washington backed the allied government of South Vietnam. When President Lyndon B. Johnson began defending that beleaguered country with combat troops in 1964, most Americans supported what they deemed a noble war against global communism and an easy fight against Ho Chi Minh's technically primitive North Vietnam and its guerrilla allies in South Vietnam.[11] That support fell off several years later as America's détente with the Soviet Union and diplomacy with China indicated that the United States was mixed up in a Vietnamese civil war rather than fighting against an expansionist and global communist monolith. Public support further declined as America's military advantages failed to translate into victory in Vietnam. The conventional calculus of overwhelming military force did not apply to this unconventional war against unyielding guerrilla forces. Although it deployed fearsome firepower, the world's foremost military failed to stop a dogged insurgency even after spending many billions of dollars and sacrificing tens of thousands of American soldiers. President Nixon began extricating an exhausted nation from this unsustainable commitment in 1969 by conducting US military operations from the air while transferring the deadly ground war back to the South Vietnamese army. This so-called Vietnamization of the war was part of the broader Nixon Doctrine, a foreign policy shift in which an overextended United States conducted what one historian called "containment-on-the-cheap" by boosting the aid and weapon sales its allies needed to carry on their own fights.[12] Little more than two years after the United States handed the whole war back to South Vietnam in early 1973, enemy fighters toppled that government and unified the country as the Socialist Republic of Vietnam.

The Vietnam War reduced America's capacity to exert hegemonic influence around the world. Its often brutal conduct on the battlefield sullied the reputation for benevolent works that facilitated foreign support for its international policies. More importantly, its failure to win that war laid bare the limits of US military might and weakened its ability to coerce reluctant nations to lend such support. The domestic consensus that had fortified American internationalism unraveled at the same time as so-called neoconservatives wanted to revive a muscular containment policy, liberal centrists pushed détente and diplomacy, and leftist critics decried violent militarism and US foreign policy as imperialism. Americans of most political stripes welcomed the end of universal male conscription after the Vietnam War, and many remained suspicious of costly US military adventures for years to come. This was true not only of peaceniks and liberal proponents of diplomacy. The sobering lessons of Vietnam also haunted military officers and conservative pols. Many determined to avoid future quagmires by deploying combat troops only when the nation's interests were clearly at stake and only with the public support and force levels needed to guarantee victory. Finally, the Vietnam War diminished America's position in the world by contributing to the sharp decline in an economy that had towered over all others since the Second World War. The United States had become the world's dominant economic power during that war. It experienced three subsequent decades of growth in which real per capita income doubled and the size of the economy swelled 140 percent. Several short recessions aside, Americans typically enjoyed robust employment and low inflation during these years. With median wages rising and employee benefits expanding, particularly for members of a historically large labor union movement, a burgeoning middle class heartily consumed the growing bounty of America's farms and factories. As US colleges opened their doors to millions of new students, an increasingly educated polity and skilled workforce found jobs in ever more productive enterprises. Private investment, high productivity, and ample consumer demand fueled this expansion, but so too did federal spending. Washington subsidized farmers and promoted agricultural modernization. It encouraged innovation among the industries it contracted for military and civilian R&D. Government also pumped money into universities through veterans' scholarships, graduate fellowships, and grants for faculty research. By launching the interstate highway system in the late 1950s, it spawned a modern transportation network that boosted auto sales and construction across the country. Every year more than a million upwardly mobile workers took advantage of government-backed mortgages and homeowner tax benefits and moved into brand new suburban houses built near these sprawling highways. Although more than one-fifth of the nation could not afford this residential version of the American Dream in 1960, Presidents John Kennedy and Lyndon Johnson used enhanced federal revenues from a growing economy to extend some of the benefits of an affluent society to these impoverished Americans. Their costly social welfare programs helped reduce poverty by nearly 50 percent over the course of the decade, leading *Time* magazine to predict in 1967 that the United States would soon "no doubt, write finis to poverty."[13]

The federal government also played an important economic role by facilitating international trade. At the end of World War II, the United States had well over half of the world's productive capacity. Americans wanted overseas markets for their surplus goods, investors looked for foreign opportunities to turn a tidy profit, and war-torn countries in Europe and Asia needed those products and capital for reconstruction. Washington initially made this exchange possible by extending the money that impoverished allies required to buy American products and kick-start their devastated economies. Its longer term plan for international commerce entailed negotiating freer trade and investment pacts with foreign partners and pegging their currencies to a US dollar redeemable for gold. Once these countries had exchangeable currencies with relative values stabilized by International Monetary Fund loans, and once Third World nations could finance infrastructure development by borrowing money from the new World Bank, international trade and investment grew mightily. Although America's relative share of global production dropped off in the 1950s, it remained the world's largest producer by far and usually maintained sizable trade and account surpluses. These surpluses not only reflected the disproportionate strength of the US economy, but they also constituted a persistent weakness in this international trading system. The many countries whose imports from America exceeded exports to it regularly experienced a "dollar-gap crisis"; they often had inadequate reserves of dollars to finance further trade with the United States. The short-term solutions of borrowing dollars or devaluing their currencies relative to the greenback did little in the short term to diminish America's economic advantages.

By the late 1960s, however, these structural trade imbalances changed as new and highly productive industries in economically resurgent Europe and Japan competed effectively with American firms. Potential exports were also lost as many US corporations built factories in other countries to take advantage of lower labor costs or assure access to foreign markets weakened by the dollar-gap crisis and protected by import tariffs. As dollars flowed abroad to service these foreign investments, finance imports, and pay for US aid and the war in Vietnam, America's balance of payments weakened and forced it to take drastic financial measures. President Johnson's devaluation of the dollar in 1968 reflected the eroding dominance of the US economy and exacerbated the domestic inflation that then persisted for more than a decade. A combination of high consumer spending and mounting government outlays for "guns and butter," namely the costly Vietnam War and Johnson's expanded social welfare programs, had pushed prices for goods and services up significantly. President Nixon inherited this overheated economy and tried to stabilize it by tightening federal expenditures and imposing temporary economic controls in 1971. But a reciprocal spiral of rising wages and mounting prices continued thereafter and fueled an unusual mix of elevated inflation and unemployment known as "stagflation." When the Organization of Petroleum Exporting Countries (OPEC) embargoed oil exports in late 1973 in retaliation for US military support for Israel during its war with Egypt and Syria, this economic downturn became even more painful as gasoline prices soared through the roof. They remained high even after the short-lived embargo, and they spiked again in 1979 after Islamist revolutionaries hostile to the United States assumed power in oil-rich Iran.

This shocking leap in energy prices cut to the heart of America's economic and political weaknesses. After three decades of impressive growth and dominance over international markets, the United States hit a wall in 1973 as its economy virtually stalled. As low-paying service work continued to expand while union labor declined and workforce productivity flattened, household incomes stagnated even as more families relied on multiple wage earners to make ends meet. Pinched by sluggish wages, sky-high interest rates, and rising prices for necessities such as gasoline, Americans were understandably gloomy about their economic prospects.[14] They may have welcomed the influx of small foreign cars after the oil embargo for their low cost and fuel efficiency, but Japanese and European imports compounded that pessimism by battering American automakers and tipping the US trade balance further into the red. OPEC's actions inflamed Americans' economic woes and despair over their nation's waning world power. The superpower that had defeated fascist empires and largely contained global communism now seemed impotent before restive Third World countries. The humiliating limits of US power became painfully evident that year when it gave up its long struggle in Vietnam and failed to prevent petroleum exporters from significantly raising oil prices. Little wonder then that Vice President Rockefeller recognized 1973 as an ominous turning point for the United States, which had "emerged from World War II as the strongest country in the world, economically and militarily." Higher energy prices signaled its dwindling strength and revealed what Rockefeller saw as a further challenge to America's economic growth and world leadership, "the limited amounts" of critical natural resources "currently available for cheap and easy exploitation."[15]

Malthusian prognosticators had previously warned of impending shortages of food and minerals critical to modern industry. In his 1948 best seller *Our Plundered Planet*, conservationist Fairfield Osborn predicted rising international discord due to growing populations and diminishing productive lands. That same year William Vogt's *Road to Survival* argued that industrial civilization would collapse if wasteful exploitation of scarce natural resources continued. These widely read books and even a 1957 NSF report that warned that "national survival may depend" on the discovery of new resources failed to darken the generally Pollyanna mood of an increasingly prosperous society. Americans had generally been optimistic that with their impressive science and technology they faced, in the words of one historian, "no limits to progress."[16] Over the course of the next decade, however, that optimism flagged as Americans opened their eyes to the dark underside of their science and technology, particularly the apocalyptic dangers of nuclear weapons and the environmental and public health costs of their industrial lifestyle. Those costs were plainly visible in the contaminated environment of the United States and in Southeast Asia, where America's high-tech weapons laid terrible waste to the land and people of Vietnam. Since these costs became evident as the US economy stagnated, foreign competitors emerged, and a growing bloc of young countries demanded greater economic autonomy, many pundits gave up their dreams of engineering global peace and plenty. They worried that the American Century had devolved into a polluted zero-sum world of tooth-and-nail competition over scarce resources.

A wave of academic works captured this gloomy mood and recapitulated the arguments of Osborn and Vogt. In his 1968 blockbuster *The Population Bomb*, biologist Paul Ehrlich declared that unchecked global population growth would trigger mass starvation, environmental deterioration, and war. American agronomists were then exporting a "green revolution" of modern farm practices and high-yield seeds that later increased Third World food production significantly, but Ehrlich's grim future seemed nigh in the early 1970s when drought and poor harvests battered food stocks around the world. Owing to the consequent run on US farm products and rise in domestic food prices, a blue-ribbon commission acknowledged that: "With such a population explosion, some people doubt whether a widespread food crisis can be averted." It determined that: "If such a crisis does occur, the people of the United States cannot expect to escape its deep impact."[17] Another stream of doomsday books looked not at population growth but at environmental degradation as the primary source of a gathering global crisis. Barry Commoner's *The Closing Circle* raised this alarm in 1971 and warned that the "flawed" technologies that recently engendered agricultural and industrial prosperity also doomed people's long-term prospects by befouling their natural environment.[18] One of the most influential works at the time, the 1972 *The Limits to Growth*, determined that the dire "predicament of mankind" stemmed not only from population growth and pollution but also its unsustainable agriculture, industrial production, and natural resource extraction. If these five trends continued, the report concluded, "the limits to growth on this planet will be reached sometime within the next one hundred years."[19]

The report was a sensation and contributed to a national debate that lasted throughout the decade. Backed by computer-assisted statistical analysis and produced by an impressive team of international specialists, *The Limits to Growth* carried weight among Americans who worried that pollution and rising commodity prices heralded a new age of economic stagnation and intense competition over natural resources, particularly fossil fuels. Respected authorities often stoked their fears. The National Science Board, for instance, declared in 1974 that "worldwide exhaustion of natural gas may be anticipated in this century, and of oil early in the next century."[20] Scientists meeting at the Fridtjof Nansen Foundation in 1975 were more specific and announced that "at the current rate of offtake there is enough oil now known and proved to last 30 years."[21] Americans needed to come to terms not only with this "unsettling energy picture," the director of the NSF warned that same year, but also with environmental degradation and "real and projected scarcities of resources and food."[22] The Global 2000 Report to the President of the U.S. (1980) sounded most like The Limits to Growth when it warned that: "If present trends continue, the world in 2000 will be more crowded, more polluted, less stable ecologically, and more vulnerable to disruption that the world we live in now."[23]

This neo-Malthusian retreat from the core assumption of the American Century, namely that the United States and its Free World partners would enjoy ever-expanding peace and prosperity, had its vocal critics. Herman Kahn of the conservative Hudson Institute rejected what he called a mistaken view "among many scholars and journalists that a turning point has been reached" portending "a much more disciplined and austere—even bleak—future for mankind." Having achieved notoriety for insisting that even nuclear warfare would not necessarily doom humankind to such a bleak future, Kahn insisted that American R&D would overcome any hurdle to progress and envisioned a far more sanguine "scenario for a 'growth' world that leads not to disaster but to prosperity and plenty."[24] Vice President Rockefeller agreed that US science and technology were necessary tools for dealing with Americans' "unprecedented challenges—in energy, in health, in food production, and in the enhancement of our environment." But he conceded that such faith "in growth and progress and in the importance of America's technological leadership seems to some to be out-of-date, old fashioned, or just plain wrong."[25]

Americans' belief that they could achieve perpetual progress by pioneering the boundless frontiers of science and technology had indeed fallen out of fashion. What Robert Heilbroner discerned in 1974 as their waning "sense of élan and purpose" and faith in "the engineering of social change" was due to their much discussed failure to protect the natural environment, vitalize America's economy, and shore up its international standing. These lapses were compounded by what he perceived as a "civilizational malaise," a spreading sentiment that modern society and its emphasis on scientific and technological progress had also failed "to satisfy the human spirit."[26] Many leading lights worried that malaise had aroused public animus to technology and the scientific enterprise. Famed rocket designer Wernher von Braun railed in 1971 against what he perceived as a "climate of irrational hostility that seems to be growing in this country," especially among humanist scholars but also the public at large, "regarding science and technology."[27] What one university president later slammed as "a massive shift of confidence away from the cognitive, or the knowing process, toward the affective, or the feeling process," the Nobel laureate I. I. Rabi subsequently admonished as a growing "anti-intellectual feeling against the use of reason as opposed to sentiment in the ordering of human affairs."[28]

Although polls conducted for the NSF showed that people's esteem for science and scientists remained high, even as their trust in politicians and other experts tanked, the anxieties of these prominent figures were not unfounded.[29] Reams of newsprint bemoaned the loss of public confidence in science during these years.[30] Many social critics impugned scientists and engineers for helping the United States become a powerful warmonger, despoiler of nature, and alienating technocracy.[31] Entertaining hagiographies of idealistic scientists also declined as Hollywood and print and broadcast media competed for more jaded and excitement-hungry audiences with darker portrayals of power-hungry scientists or reckless eggheads who unwittingly released malign forces on the world.[32] Scientists felt the additional sting of academic scrutiny as sociologists and historians knocked them off their gilded pedestals and studied them as professionals rather than special devotees of natural truths. Rejecting positivist epistemology, these academics showed that scientists were not always objective, forward-looking students of the natural world or true to what one NSF chief had hailed as the "apolitical and 'self-purifying'" character of science.[33] Rather they were mere intellectual mortals, in scholar Thomas Kuhn's famous analysis, who often resisted evidentiary challenges to their common held conceptual paradigms.[34]

These soft rebukes may have blemished the sterling authority of scientists and engineers. But starry-eyed accounts of boundless scientific and technological progress so common in the 1950s and early 1960s fell out of fashion owing more to a sudden reordering of government priorities and drop in federal R&D funding. The need for scientific personnel and the demands of a high-tech economy and national security in the decade after Sputnik had prompted Washington to increase fourfold its already substantial support for military and civilian R&D. The quixotic boosterism that had helped agencies such as the NSF and NASA justify that ample funding dropped off as shrinking government outlays went increasingly to mundane, practical R&D. After nearly 20 years of steady growth, federal funds declined after 1967 as an earlier scarcity of scientists and engineers turned into a glut and a tightening economy could no longer bear this exorbitant R&D. Détente with the Soviet Union, diplomacy with China, and increased arms sales to its allies simultaneously tempered the need for costly US defense expenditures. As executive and legislative leaders turned their attention away from Cold War confrontation and to mounting social and economic problems, they encouraged more applied research and expected recipients of their shrinking R&D outlays to address pressing national concerns such as environmental pollution, energy insecurity, urban decay, and falling economic competitiveness. Dealing with these mundane issues was certainly a matter of national urgency, but doing so signaled the onset of narrower and more practical expectations about science and technology.[35]

A glimmer of this gathering utilitarian ethos came in 1966 when a Department of Defense (DOD) study rejected the core principle of Vannevar Bush's seminal 1945 policy report *Science: The Endless Frontier*. Bush insisted that ample support for basic research was critical to America's military and economic security since science generated practical knowledge essential for technology development. The DOD study dismissed this orthodoxy and determined that "technological achievements stemmed primarily from mission-oriented engineering research and development" rather than fundamental science. The NSF, created in 1950 to be America's primary sponsor of basic research, predictably defended Bush's position in 1968 even as it fell under the sway of practical demands. While the NSF's budget grew to cover basic science no longer funded by other mission-oriented federal agencies, it also followed a new Congressional charge to sponsor more practical research through novel initiatives such as its Research Applied to National Needs Program (RANN). This drop in federal support and shift toward applied R&D led the editor of the *Bulletin of the Atomic Scientists* to warn that the "scientific revolution of the 1950s and 1960s has given way to a counter-revolution which deems pure science irrelevant if not inimical to the real concerns of society."[36] A political consultant working with the NSF agreed and privately counseled agency officials in 1976 that because *Science: The Endless Frontier* was "being parodied in Congress as 'Science, The Endless Expenditure,'" scientific institutions needed to "look for and cultivate major support from the public at large in order to achieve a public base of support."[37] The way to do so, of course, was to promote the direct practical spinoffs of scientific research. At a time when popular culture caricatured scientists, academics scrutinized their professions, and federal patrons applied more strings to their reduced funding, the National Science Board sullenly reported in *Science at the Bicentennial* that the "last 10 years have shaken the confidence of scientists and engineers."[38]

What historian James Patterson called the "exciting and extraordinarily expectant thirty years following World War II," when fantastic advances in science and technology validated talk of conquering new frontiers, seemed to end in the early 1970s.[39] The "public institutions and the social values that had accounted for the astounding progress of the United States," one presidential commission wrote at the end of the decade, "were now straining to cope with the massive problems of the era." It regretfully concluded that the halcyon days when "we believed that we could do almost anything we set out to do" had passed owing to "a decline in public support for science and technology, closely related to expectations of material progress that seem more difficult to satisfy and to fragmentation of the American public."[40] Such public anxiety about the plummeting fortunes of a socially fragmented nation had also been evident nearly a century earlier when such leading lights as historian Frederick Jackson Turner and future president Theodore Roosevelt pinned America's mounting challenges on the closure of its continental frontiers. Whereas Turner nostalgically bemoaned the loss of those frontiers, Roosevelt saw them as a still vital model for societal uplift. In his estimation Americans could overcome their enervating modernity and maintain their great nation's march forward by adopting the manly independence and martial bearing of their pioneering predecessors. Americans suffering "civilizational malaise" in the 1970s were theoretically ripe for Roosevelt's revitalizing brand of frontier nationality. They had spoken confidently for many years about the nation's new frontiers, and many yearned to carry on its pioneering achievements and world leadership during its third century of existence. But that invigorating discourse did not endure, for the racial, gender, and martial sentiments at the heart of Roosevelt's frontier mythology no longer suited Bicentennial America.

Despite their history of racial discrimination, Americans had long exalted their nation as the embodiment of universal values, an inclusive society based on "the fundamental equality of all human beings." This tradition of "civic nationalism" coincided with a very different "ideological inheritance," what historian Gary Gerstle calls "a racial nationalism that conceives of America in ethnoracial terms, as a people held together by common blood and skin color and by an inherited fitness for self-government."[41] Theodore Roosevelt embraced the civic ideal that the United States was a melting pot that incorporated

diverse peoples. But he also felt that their unique freedom and prosperity had been forged by a racial elite who put their superior Teutonic
inheritances to work under the strenuous conditions of a pioneering
society. During the first two decades of the Cold War, Roosevelt's
bridge between these civic and racial traditions tenuously held. For
example, the Americans who plied the endless frontier of science and
the distant frontiers of Antarctica and outer space appeared to be representatives of a great civic nation struggling to make the world free
and more prosperous. These envoys of liberty and prosperity were
also primarily white men, and their celebrated exploits reinforced the
still common belief that America's highest achievements came from
the uniquely capable hands of its manly white vanguard.

This tenuous racial balance shattered over the course of the 1960s
as race discrimination damaged America's reputation and a growing civil rights movement struggled heroically to secure equal rights
and opportunities for all people. If that movement's increasingly
mainstream vision of an integrated, color-blind society doomed the
Rooseveltian conceit that white privilege and democracy could coex-
ist, so too did multiplying racial and ethnic splits in America. Racial
injustice endured even as US courts and Congress repealed discrimi-
natory laws during the 1950s and 1960s. Widespread urban turmoil in
that latter decade accelerated white suburban flight, and some of the
millions of African Americans consigned to urban ghettoes embraced
Afrocentric movements that preferred black traditions and communities over social integration. The simultaneous revival of ethnic pride
among Native American, Hispanic, and white ethnic groups, combined
with the shift among advertisers and campaigners from mass markets
to segmented audiences and interest group politics, created a complex
and fragmented cultural landscape. Roosevelt's pioneer hero enjoyed
wide currency when the public sphere was simpler and dominated by
members of his preferred races. But as their hegemonic influence over
a fragmenting culture slipped, Roosevelt's racially chauvinistic frontier
myth became less popular and lost its mass audience.[42]

The racist tilt of that now-discredited myth negatively affected
the trope of the space frontier, pioneered till then by widely admired
white astronauts. Since race had been less evident in public accounts
of US Antarctic exploration, the Rooseveltian motif of the Antarctic
frontier suffered more from a similarly profound change in American
gender relations. A still patriarchal society had shared Theodore
Roosevelt's belief that men alone had the strength and grit to tackle
forbidding frontiers like the southern continent. This attitude was
evident when a *National Geographic* journalist joked in 1957 that

Antarctic weather was "as changeable as a woman's mind" and a NSF official praised each south polar scientist in 1968 for being "man enough to leave the comforts of his wife's bed."[43] By the end of the 1960s, such overt gender chauvinism receded from the public sphere as female employment expanded, a women's movement pushed effectively for equal access to education and employment, and feminist activists challenged deep-rooted beliefs about women's "natural" virtue and domesticity. Although an *Audubon* essayist bucked this trend in 1973 when he described Antarctica as the "one virgin continent—unviolated because of its ring of ice that protects it like a chastity belt," such sexist metaphors were rare by that time.[44] Women's incipient participation in the USAP had already proven old stereotypes wrong. Under pressure from the NSF and new equal opportunity statutes, the Defense Department in 1969 finally accommodated women scientists who wanted to work with the USAP, and the Navy recruited female ensigns to the continent in the ensuing years.[45] While these women did not embody the manly heroism associated with Antarctic exploration, nor did they signify what Roosevelt would have regarded as an advanced stage of frontier domestication. They went to Antarctica not as dainty homemakers but as intrepid researchers and naval recruits who successfully worked alongside men in a still rough land.

Finally, Roosevelt's frontier mythos lost its cultural salience because its story of national progress through Indian wars and land conquest no longer appealed to Americans who had turned away from such bellicose action. The Vietnam War not only coincided with a steady increase in violent content in American popular culture, but it also triggered the end of universal military service and a decline in the public's ritualized devotion to war. What one cultural observer identified as a centuries-old myth of a nation triumphantly forged through noble wars virtually disappeared. Many Hollywood movies, for instance, turned once heroic soldiers into tragic victims or sometimes unhinged perpetrators of national folly.[46] This was the case not only in Vietnam War films but also in long popular Westerns, which began inverting the traditional frontier narrative by maligning cowboys and cavalry as trigger-happy land grabbers and ennobling Indians as peaceable defenders of family, home, and nature.[47] While this desacralization of war stripped away a critical element of American frontier nationality, the concurrent eruption of environmental concern was the primary reason the Rooseveltian trope of the Antarctic frontier collapsed. Presidential cables and visiting journalists alike still applauded the now coed Antarctic personnel for their courage and sacrifice, but they no longer cast them as pioneer conquerors of a savage continent. This martial account of environmental subjugation no longer sat well among the "millions of our citizens" whom President Ford said "share a new vision of the future in which natural systems can be protected, pollution can be controlled, and our natural heritage will be preserved."[48] Spirited talk of an Antarctic frontier faded fast as these new environmentalists discovered the southern continent and questioned their earlier emphasis, in the words of one scholar, on the "conquest and development of nature."[49]

Thus Americans may have yearned for a revitalizing narrative of national character and purpose, but public officials and opinion makers avoided a Rooseveltian frontier myth that had fallen out of synch with their fast-changing culture. Proponents of the USAP initially responded to this change by simply touting the mundane benefits of south polar research. Their rhetoric was viable in the short term. While they had used the motif of the Antarctic frontier to excite public support for a robust program essential to US prestige and position in the region, post-Sputnik pressure to demonstrate America's scientific and technological superiority had dramatically abated. Whereas America's regional interests had depended on its clearly preeminent footprint in Antarctica, they were now tenuously secured by the Antarctic Treaty system, which had prevented skirmishes from breaking out there between the superpowers and among US allies since 1959. Their downsized talk of basic science and amicable Antarctic relations, however, were not sustainable in the long run. Attentive media outlets and proliferating science museums naturally sought more compelling justifications for Antarctic exploration.[50] Policymakers did so as well when new challenges to US regional interests emerged within the Antarctic Treaty system and among excluded nations and nongovernmental organizations critical of its management of the southern continent. Although US officials and pundits avoided the now-discordant strand of frontier nationality associated with Theodore Roosevelt, they flirted with the vein linked to Frederick Jackson Turner by predicting that Antarctica would soon become the world's safety value, the resource-rich frontier it needed to overcome what many feared were global limits to economic growth.

Resource Development and Environmental Protection

Frederick Jackson Turner agreed with his contemporary Theodore Roosevelt that America's exceptional character stemmed from its history of frontier conquest. Whereas Roosevelt was concerned with racial and societal vitality and emphasized the contemporary relevance of that martial endeavor, Turner regarded it nostalgically as a lost source of socioeconomic progress. That hinterland had not only stimulated

Americans' unique values and political system, he concluded, it also saved the United States from the sorry fate of geographically bounded nations. Without empty land to absorb growing populations and host emerging markets, such societies suffered economic volatility and class conflict. The pioneering United States had avoided this fate because surplus laborers moved to the frontier and earned income to purchase the nation's prodigious manufactures. They also created profitable opportunities for metropolitan investors and provided bountiful food and critical raw materials to America's voracious industries. In short, Turner believed that vacant lands had made for a free, democratic, and prosperous nation. Having eulogized the closing of America's landed frontiers in 1893, Turner looked for and found their alternative in the science and engineering labs of America's public universities. He believed that these institutions were open to enterprising men of talent whose research and technological innovations would create novel industries and turn the diminishing wealth of exploited lands into new bonanzas of useful natural resources.

In the decade after the 1957–58 IGY, public accounts of Antarctica as a new frontier and the USAP as a pioneering endeavor contained these bread-and-butter elements of Turner's frontier thesis. It had been commonsense for decades that the untouched Antarctic, like the putatively empty lands of America's western frontiers, would eventually provide Americans with valuable natural resources. Furthermore, since high-tech scientific exploration of Antarctica was akin to Turner's alternative frontiers of science and technology, observers naturally expected American explorers to discover ways to make the frozen continent yield its hidden riches; so predicted Metro Goldwyn Mayer's 1948 movie *The Secret Land*, which called US naval expeditions there an important step in exploiting the "one untouched reservoir of raw materials left in the world." Until his death in 1957, the celebrated polar explorer Rear Admiral Richard Byrd similarly waxed about US scientific exploration in that "untouched reservoir of natural resources."[51] In the ensuing years, as Americans worried about their nation's post-Sputnik standing and what one distinguished study group called "the problem of national purpose," proponents treated the USAP and the prospect of harvesting Antarctic resources as an answer to that group's most pressing question: "what the United States ought to do with its greatness."[52] They were confident that Antarctica and every other corner of the planet would soon submit to America's grand project to engineer the world so that even the weather was brought to heal. In this "era of unlimited power," a 1968 documentary on Antarctica accordingly explained, the US would soon make use of that continent's hidden wealth and even "change the temperature balance and convert the cold regions to productive areas."[53]

As America's economy and international standing faltered in the early 1970s, many people asked a very different if even more pressing question: what had happened to the nation's greatness? In this era of diminished expectations, public officials and media observers recalibrated their pitch about Antarctica even as they remained hopeful about south polar resources. They started speaking more guardedly about humankind's desperate needs rather than a visionary project of a great nation. But their final message remained the same; Antarctica would soon provide the world with essential resources. Several scientists pointed to Antarctica's unique geography as a resource that might save industrial civilization from the limits to growth. Undeterred by the 1959 Antarctic Treaty's prohibition on nuclear disposal, they identified the southern continent as the only viable location to deposit radioactive wastes from nuclear power plants. They believed that atomic energy alone could fulfill the world's growing demand for electrical power and that people needed to bury those dangerous wastes a world away under the more than two-mile thick Antarctic icecap.[54] Others looked to the continent's most abundant resource, its seemingly limitless expanse of ice, as a saving grace. Towing Antarctic icebergs to irrigate the economically booming deserts of southern California and the Middle East may have been, in the words of *The Sacramento Bee* newspaper, a pie-in-the-sky "Jules Verne-like scheme." But the California legislature endorsed the idea in 1977, as did a state representative who formally raised the issue on Capitol Hill. The academics who gathered at Iowa State University the next year for a conference on *Iceberg Utilization* garnered few headlines, but Saudi Prince Mohammed Al-Faisel attracted favorable media attention when he formed Iceberg Transport International Limited, a short-lived company meant to tug Antarctic icebergs to his oil-rich but water-starved country.[55]

With public attention then fixed on rising food and energy prices, Antarctica's marine and mineral resources stood out as its most valued assets. Whereas sealers and whalers had once monopolized its teeming waters, commercial fishing vessels moved into the region at this time. Walter Sullivan of *The New York Times* reported in 1960 that ships had begun harvesting Antarctic fish and zooplankton, which a 1969 US Navy report had in mind when it declared: "With the rapid growth of population in all parts of the world, every source of food will be needed, and the abundant life of the southern seas may contribute notably to the future of mankind."[56] Biologists working in the region since the IGY had turned their attention to krill, a tiny plankton-feeding invertebrate whose massive swarms had already attracted Soviet fisher-man. The scientists who sat on the international Scientific Committee on Antarctic Research (SCAR) were also impressed by krill, which they deemed Antarctica's most likely bonanza of food for an explod-ing human population. Influenced by SCAR's findings, *National Geographic* summarily reported in 1968 that "the virtually limitless shrimp-like krill, a vital link in the food chain of Antarctic waters, may provide a huge resource for a hungry world."[57]

While a decade of offshore research revealed the vast proportions of Antarctic plankton, krill and fish, mounting evidence attracted world attention to its likely mineral and oil resources. According to an NSF documentary, geological findings in Antarctica since the IGY turned the "vague idea" of continental drift "into an accepted theory."[58] Antarctic fossils of long extinct plants and animals rein-forced that theory by indicating that several continents had been clustered together in what earth scientists called "Gondwanaland" before drifting apart over the past 200 million years. Many special-ists regarded the 1969 discovery of Lystrosaurus fossils in Antarctica, which were also located in South America and Africa, as confirming evidence for this theory and the mechanism of plate tectonics, since this small land reptile could not have crossed the oceans now sepa-rating these ancient Gondwana neighbors.[59] Many people logically assumed that if Antarctica was once connected to those other land-masses and hosted the same flora and fauna, then it also had similar mineral deposits and petroleum reserves hidden beneath its thick ice and deep continental shelf.[60]

This prospect piqued the attention of several governments and petroleum companies, which began eyeing Antarctica as a likely fount of energy resources. It certainly generated front page headlines in the United States and a buzz in the secretive Kremlin, which the Central Intelligence Agency believed had made "the discovery of min-eral resources...an important component of each Soviet Antarctic Expedition."[61] That interest picked up after the OPEC embargo, as high oil prices turned marginal and hard to reach deposits into poten-tially profitable fields, and as news leaked out that the US Geological Survey (USGS) research vessel *Glomar Challenger* discovered traces of hydrocarbons off the Antarctic coast in 1973. As one scholar later wrote, this news of possible "mineral riches in this wholly untouched continent stirred the most fabulous hopes."[62] Whereas knowledgeable journalists like Walter Sullivan had recently assumed that Antarctica had a "reservoir of mineral resources that will probably remain untouched for a long time," many reported after the USGS survey that resource development was fast approaching and estimated that the region "could have potential resources of 45 billion barrels of oil and 115 trillion cubic feet of natural gas."[63] These estimates were comforting indeed to people worried about an impending depletion of world energy reserves. Even though a 1976 Congressional report cautioned that enormous hurdles remained before any existing fossil fuels could be tapped there, a State Department specialist noted that oil companies and federal officials had taken keen note of these potential energy resources. So too had the American public, whose "flurry of interest in mineral resources, especially petroleum, stemmed from the period of the 1973 OPEC oil embargo and the coincidental reports of possible oil reserves offshore of the Antarctic continent."[64]

That flurry of interest came in a surge of public statements about the USAP and of news stories about the tantalizing potential of Antarctic resources. Sensitive to their readers' concerns about energy and demand for practical R&D, journalists made resource development a central feature of their accounts of south polar exploration. Some public officials welcomed US development of an Antarctic oil frontier as a fitting national project. But Washington did not spring immediately into action.[65] Petroleum had not yet been found there, and it would have been difficult and expensive to tap. More importantly, State Department experts felt that an oil rush would politically destabilize the region, while NSF officials worried it would hurt America's important program of basic science in Antarctica.[66] NSF Director H. Guyford Stever acknowledged the pressing need for directed research on resource development, which was a priority for his agency's RANN program. Stever also accommodated new federal Office of Management and Budget (OMB) guidelines that compelled the NSF to detail how its Antarctic program targeted the area "as a resource base." But another OMB criterion emphasized how USAP funding supported research on the region "as an environmental benchmark."[67] Although environmental research in Antarctica did not promise the economic windfall of oil development, proponents usually described such research in practical rather than romantic terms. Environmentalists would soon advocate preserving Antarctica as a sublime wilderness, but utilitarian talk of research in a unique and protected natural laboratory was then the primary alternative to Turnerian calls for exploiting a resource-rich frontier.

Such instrumental environmentalism had long existed in the United States. Some of the nation's earliest and most celebrated environmentalists, such as Henry David Thoreau and John Muir, had romantically venerated wild nature as a sublime source of moral uplift. But their environmentally conscious contemporaries were generally more concerned with material ends. Some late nineteenth-century reformers, precursors to modern environmentalists, combated factory pollution and urban refuse as costly burdens to urban society, while resource-minded "conservationists" at the time tried to rationally manage America's natural capital. President Theodore Roosevelt lionized backwoodsmen who cut wide swaths across the nation's frontiers, but he was also an ardent conservationist who believed that it was then essential to check such unregulated action and develop America's natural resources in a sustainable manner. Many of the professionals who staffed the state and federal agencies tasked with stewarding the public domain similarly wished to avoid all manner of environmental havoc triggered by haphazard resource exploitation. The primary focus of these conservationists was the practical use of natural resources.

This conservationist impulse was evident among many Antarctic scientists after the IGY. The American ornithologist Robert Cushman Murphy argued in 1962 that Antarctica should be protected since it was the only remaining area that "man has not yet occupied, saturated, and extensively changed." He proposed designating Antarctica an "international park" with a practical goal in mind, conserving for the long-term benefit of natural science what many experts believed was a precious rare phenomenon. They regarded Antarctica as the world's largest "primitive climax," a whole continent whose natural system had matured to a final evolutionary state.[68] The Antarctic Treaty nations did not follow his advice, but they too saw the scientific utility of an undisturbed icescape and agreed in 1964 to more limited measures. They cordoned off environmentally vulnerable areas, controlled polluting discharge from ships and stations, and protected vulnerable birds and marine mammals. Treaty representatives also worried that a new hunt for sea mammals would inflame sublimated territorial conflicts and undermine important marine zoological research, so they implemented a treaty in 1972 to protect Antarctic seals.[69]

This burgeoning conservationism was evident by 1971 when the *Antarctic Journal of the United States* highlighted for the first time the NSF's efforts to protect the fragile continent. One article in particular demonstrated how dramatically public attitudes about nature and the Antarctic environment had changed. When naval Seabees built an IGY research station at Cape Hallett in 1956, a *National Geographic* journalist and NBC television crew gave comical, firsthand accounts of the messy removal of nearly 8,000 nesting birds there. Whereas the magazine blithely described the operation and NBC jokingly called it the "greatest population relocation since Longfellow's 'Evangeline,'" the 1971 article frankly condemned the destruction of wildlife habitat at Cape Hallett and described the NSF's efforts to clean the site and repopulate the rookery.[70] Fifteen years after the station was built, the NSF considered the removal of penguins not as a humorous sideshow to an important scientific enterprise but as a blow to the environment and scientific utility of a pristine Antarctica.

The Turnerian resolve to exploit Antarctic resources faced a related but deeper challenge posed by modern environmentalism. According to one historian, the upswell of environmental concern in the United States and around the world by the late 1960s was based not on the old conservationist "fear of running out of resources" but from a "new driving impulse [that] transcended the concern for the quality of life to fear for life itself."[71] The stupendous technoscience advances that underwrote Americans' fantastical faith in what one observer called their "superhuman abilities to control the physical and chemical attributes of nature" also allowed them to expect healthful and pleasing environments.[72] The friction between these countervailing tendencies first became evident in the late 1950s with regards to nuclear power. Government and industry boosters had tapped that faith in superhuman ingenuity by promising a world positively transformed by civilian applications of atomic energy. But this gathering environmental entitlement spawned new and widespread denunciations of atomic power in the late 1950s. Americans grew concerned not only about the destructive power of nuclear weapons, but also about environmental contamination from radioactive fallout associated with atmospheric testing of nuclear bombs. They discovered that fallout from these tests circulated throughout the world and percolated in greater concentrations up the food chain and into human bodies. When radiation from these tests appeared in mothers' breast milk and babies' teeth, international opprobrium mounted and compelled the superpowers to ban such tests in 1963. By then Americans had awakened to a novel crisis, a planet they had dangerously polluted. Many scholars credit the biologist and award-winning nature writer Rachel Carson with turning this nascent environmentalism into a full-scale national passion. By tapping people's fear of fallout and persuasively arguing that aggressive use of pesticides such as DDT similarly contaminated the environment, bioaccumulated up the food chain, and threatened human health, her 1962 blockbuster *Silent Spring* made Americans widely aware that the tools and habits of industrial civilization threatened the fragile fabric of nature. National media gave front-page attention to environmental pollution after Carson's book, while a growing number of environmental organizations lobbied to protect the natural landscape. Whereas nature-friendly politicians and civil servants had recently concentrated on establishing parklands and wildlife preserves, by the late 1960s they endeavored to stem the dangerous proliferation of environmental pollutants. To do so, the NSF director explained, "we must look to science" not only for material progress but "for the skills and wisdom that will enable us to bring man into harmony with his environment."[73]

Informed by what one historian called this "science-based sense of relatedness between man and nature," people the world over mobilized to deal with a host of environmental threats.[74] Membership in environmental organizations in the United States doubled in the early 1970s, and doubled again over the following decade. By the time millions of Americans expressed their mounting concerns for nature during the first national Earth Day celebrations on April 22, 1970, the US government had assumed a greater responsibility for environmental protection. It had registered an unprecedented burst of environmental legislation and created, later that year, the federal Environmental Protection Agency and the White House Council on Environmental Quality. The international community took up the issues of resource conservation and nature protection as well, since many environmental problems transcended national boundaries. The 1968 UN "Biosphere Conference" stimulated such efforts and offered scientific evidence that modern agriculture and industry adversely affected wildlife around the world. That gathering set the stage for the 1972 UN Conference on the Human Environment in Stockholm, a watershed event that reoriented international attention from "the limited aims of nature protection and natural resource conservation to the more comprehensive view of human mismanagement of the biosphere." As one student of the Stockholm Conference noted, it also "marked the beginning of a new and more insistent role for [non-governmental organizations] NGOs" in environmental affairs and led to the creation of the United Nations Environment Programme to coordinate environmental initiatives among that body's many agencies.[75]

This wave of state, United Nations, and NGO activity turned Antarctica's relatively undisturbed landscape into a valuable, non-extractive resource unlike any found on Turner's frontiers. As a US presidential commission pointed out in 1971, environmental protection on a global scale "would as a first essential step include a study of major pollutants of the atmosphere and the oceans."[76] Scientists had laid foundations for that very study in Antarctica since the IGY, when oceanographers first tracked the region's icy melt water deep into the northern hemisphere and researchers began plugging data from Antarctic weather stations into global meteorological databases. IGY scientists turned Antarctica into benchmark for the planet's atmosphere by measuring local traces of fallout from distant nuclear weapons tests. They were narrowly interested in tracing the movement of radioactive particles through the physical environment and food chain. But scientists later used that data to track what they now deemed dangerous environmental pollutants. By 1965 they discovered in penguins and on the remote Antarctic icecap traces of Strontium 90 from nuclear explosions, lead from automobile and industrial emissions, and pesticides from distant farms. These contaminants vividly illustrated the global scale of industrial and agricultural pollution. "Who would have thought, even a decade ago," one exasperated scientist wrote in 1971, "that the DDT sprayed on the cotton fields of Arizona would find its way to Antarctica [and] who would have supposed that the exhaust from the automobiles of Los Angeles would pollute the last continent unspoiled by man?"[77]

The researchers who discovered these pollutants and revealed the unspoiled continent's ecological connection to distant societies paved the way to making Antarctica an international environmental cause. Owing to their experience in Antarctica, concern for its natural environment, and perhaps need to curry favor among increasingly practical-minded benefactors of American R&D, US scientists were early advocates for using the southern continent to tackle global environmental problems. Raymond Dasmann stepped in this direction in 1968 and identified the region as "a final testing ground for conservation," a place to galvanize the will and develop techniques to "solve our environmental problems."[78] Several years later biologist Bruce Parker hit on an increasingly popular idea. He proposed using Antarctica as a barometer of global pollution. Citing the presence of DDT as "the prime example of the inseverable connection between Antarctica and the rest of the world," Parker urged the Antarctic Treaty nations to avoid contaminating the local environment so that researchers there could track the global spread of industrial contaminants.[79]

The NSF recognized that an undisturbed Antarctica was an ideal platform for global environmental assessment. That research became a central priority for the USAP, and NSF officials pledged in 1971 to reduce pollution from US operations in Antarctica. They also saw Antarctic research as a key to dealing with a new and troubling environmental issue, the prospect of global climate change. IGY planners had assumed that planet-wide systems like climate would change, albeit extremely slowly, and they duly called on participating countries to make "epochal observations of slowly varying terrestrial phenomena."

Measuring those phenomena in Antarctica, including the growth or recession of its massive ice sheets, was so important that program organizers determined that "Antarctica represents a most significant portion of the earth for intensive study during the International Geophysical Year."[80] IGY surveys provided critical baseline measurements of Antarctica's ice mass, and US scientists used thousand-foot ice cores to provide the first glimpse of ancient weather conditions in the region. Because these long tubes of ice contained evidence of past atmospheric temperatures and carbon dioxide (CO_2) levels, they later helped scientists predict the climate altering effects of humankind's ongoing CO_2 emissions. IGY researchers knew that elevated CO_2 levels might raise global temperatures, but they did not yet recognize the diagnostic value of Antarctic ice cores. In any event, they figured that global warming would take at least 10,000 years to occur.[81]

Scientists shortened this timeline significantly in the early 1970s.[82] Although many then suspected the world was cooling down and heading toward another ice age, some worried that mounting CO_2 emissions might lead instead to rapid global warming.[83] They agreed that in either case the Antarctic ice cap was an important barometer of climate change and regulator of global weather. A *National Geographic* journalist who leaned toward global warming wrote in 1976 that scientists wanted to know if "the present warm interglacial climate [was] causing it to melt and raise sea levels worldwide? Or will the warmth cause added evaporation and snowfall that will eventually re-enlarge Antarctica's icecap?"[84] An NSF official hopefully speculated that the latter was true and that cooler temperatures from Antarctica might break the dry heat then hurting world farm production and "counter the southward migration of the sub-tropical deserts."[85] NSF Director H. Guyford Stever took the threat of desertification and global warming so seriously that he urged President Nixon in 1973 to finance research on the "global scale climatic fluctuations" he believed were then ruining crops around the world.[86] This issue had vexed Nixon, who worried that crop failures and associated price spikes threatened America's economy and his détente policy of selling grain to the Soviet Union. The president was not yet thinking of climate change when he praised Antarctic scientists in 1970 for "foster[ing] cooperative scientific research for the solution of worldwide and regional problems, including environmental assessment."[87] Nor was his Secretary of State William Rogers, who similarly referred to monitoring global pollution when he declared that Antarctic research had enriched "man's knowledge of his environment and his understanding of the earth."[88] However the idea that Antarctic research could shed light on climate change as well as environmental pollution soon gained traction. The editor of the *Bulletin of the Atomic Scientists* had both applications in mind in 1973 when he avowed that such research "provides essential information for the understanding of planet-wide environmental processes."[89] So too did an NSF documentary that declared the next year that scientists "know they can learn things in Antarctica that they cannot learn anywhere else. And for that reason they know it must be protected."[90]

Thus as the heroic motif of the Antarctic frontier lost ground, two alternative visions for Antarctica and the USAP emerged, one emphasizing resource exploitation and the other environmental protection and research. President Nixon felt that these two priorities were compatible and declared in 1970 that America intended to "protect the Antarctic environment and develop appropriate measure to insure the equitable and wise use of living and nonliving resources." A presidential commission took resource development for granted the next year even as it urged the United States to "take the initiative in proposing special international agreements for Antarctica in order to protect its unique environment."[91] But an early sign of the impending battle between advocates for resource exploitation and those for environmental protection appeared in 1972 when the Second World Conference on National Parks made the first concerted appeal to protect the southern continent. The Conference officially recognized the "great scientific and aesthetic value of the unaltered natural ecosystems of the Antarctic continent and the seas around it" and recommended protecting the region by strictly limiting resource extraction and designating Antarctica a "World Park, under the auspices of the United Nations."[92]

This call for environmental protection heralded profound changes in public attitudes about Antarctica and in the governance of the southern continent. Like many of their Antarctic Treaty counterparts, US officials had made Antarctic policy largely unencumbered by public opinion and interest group lobbying.[93] This began to change in the mid-1970s as US officials were tugged by opinion-makers and interest group advocates. Some pushed for resource development in Antarctica. Others called for environmental protection there. While petroleum companies had allies inside the White House and the federal Antarctic Policy Group, environmental organizations exerted the clout of their swelling memberships to appeal to Washington on behalf of the Antarctic environment. Had national interests in the region remained secure, Antarctic policymakers may have let these countervailing lobbies play out. But US officials ultimately leaned toward these NGOs and adopted a strict environmental platform when the prospect of resource development threatened the Antarctic Treaty system.

Triumph of Environmental Stewardship

The Antarctic Treaty system had engendered peaceful coexistence in the region among the United States, the Soviet Union, and 11 other signatory countries since 1959. However the Treaty remained secure only as long as a subset of those countries, the so-called claimant nations of Great Britain, New Zealand, Australia, France, Norway, Chile, and Argentina, did not exert their previous claims of exclusive sovereignty over Antarctic territory. National pride and regional influence had prompted those territorial claims, as did the attraction of Antarctica's natural resources. Since those resources were not yet economically practical or even known, the Treaty's Consultative Parties had little reason to defend outstanding land claims or assert new ones. They preferred cooperation and access to all corners of the region afforded by the Antarctic Treaty. In the late 1960s these parties worried that the emerging commercial fishery could damage Antarctica's marine food chain and awaken dormant territorial ambitions if any claimant nation wanted exclusive access to fishing grounds. They were rightly alarmed by the growing prospect of unregulated mineral resource development, which the US Central Intelligence Agency warned would "'torpedo' the Treaty by aggravating territorial claims problems" and end more than a decade of cooperation in Antarctica.[94]

Having "received enquiries from marine geophysical prospecting companies about the possibility of prospecting in Antarctic seas," the Consultative Parties formally broached the issue of mineral development in 1970. Two years later they acknowledged that oil and mineral development could turn Antarctica into "the scene of object of international discord" and "raise problems of an environmental nature." But they still thought such development was unlikely in the near term and agreed only to study the issue and take it up at their next biennial meeting.[95] By then the OPEC embargo and mounting evidence of south polar oil forced their hand. Environmental organizations urged the Treaty nations to allow the United Nations to manage Antarctica as a nature preserve, while representatives of several Non-Aligned nations agitated for UN control over the region. They wanted all countries to profit from Antarctica's presumably rich oil and mineral reserves, which they proposed designating the "Common Heritage" of all humankind. The Consultative Parties hoped in 1977 to defuse these pointed challenges to their exclusive regional influence by urging their "nationals and other States to refrain from all exploration and exploitation of Antarctic mineral resources while making progress towards the timely adoption of an agreed regime concerning Antarctic mineral resource activities." Their temporary moratorium was a pivotal moment in Antarctic affairs when the Consultative Parties began legitimizing their political control by asserting they alone could preserve the tenuous peace in the region and protect "the unique Antarctic environment."[96]

The signatory nations used environmental protection to justify the Treaty system as they tackled first the pressing problem of Antarctic fishing. They had planned to wait for SCAR and the Scientific Committee on Oceanic Research to finish surveying the region's marine life so they could set sustainable catch-limits on Antarctic fish and krill. However Treaty representatives decided to move quickly in 1977, alarmed that unregulated harvests could decimate Antarctic krill and fish just as it had recently Peru's immensely productive anchoveta fishery. They declared a "definitive regime for the Conservation of Antarctic Marine Living Resources should be concluded before the end of 1978."[97] Fishing fleets from at least eight countries were in Antarctic waters or preparing to sail for the southern ocean. Their number was likely to increase as signatories of the 1976 UN Law of the Sea Treaty took control of offshore waters far beyond the traditional three miles from their coastlines. Worried that international vessels cut off from prolific fishing grounds in these new 200-mile "Exclusive Economic Zones" (EEZs) would head to open fisheries around Antarctica, the Consultative Parties worked doggedly to institute the 1981 Convention on the Conservation of Antarctic Marine Living Resources (CCAMLR). Signatories unveiled this convention with great fanfare, despite its many weaknesses. The Convention deferred catch limits until researchers completed marine surveys, which allowed still unregulated fishing vessels to possibly strip Antarctic waters. The CCAMLR also remained neutral on the issue of territorial sovereignty, leaving signatories unprepared in the event that claimant nations asserted control over fisheries within vast EEZs extending 200 miles beyond the shores of their coveted Antarctic terrain. But these weaknesses enabled the Consultative Parties to achieve a difficult consensus and establish their environmental credentials with a path-breaking regime to conserve a whole marine "ecosystem" rather than ensure the "maximum sustainable yield" of specific species.[98] CCAMLR aimed not only to conserve krill, the region's lynchpin species. It also aimed to protect populations of Antarctic fish, marine mammals, and birds that feed on krill so that they would not be decimated by massive krill harvests.

US officials expressed sincere concern for the Antarctic environment and advocated strongly for marine conservation in the lead up to the CCAMLR. It helped that Americans did not fish Antarctic waters or lobby for greater access to them. A State Department press release accordingly declared that because "the waters surrounding Antarctica appear to be both highly productive and vulnerable to unregulated harvesting," it wanted to make sure "that utilization of living resources will take place in accord with our commitment to the Antarctic environment."[99] When US officials followed through on that commitment by inviting prominent environmentalists to sit on an Antarctic advisory committee, a State Department representative praised America's "departure from the practice of exclusion of public members prevalent as late as 1976 and still exercised by almost all other Consultative Parties."[100] Congress continued that departure by inviting prominent environmentalists to testify about Antarctica for the first time in 1978. Speaking on behalf of seven influential NGOs, the director of the Center for Law and Social Policy professed that Antarctica "plays an important role in the global natural processes that control life on earth." For this reason "all of mankind shares an interest in ensuring that the geophysical, biophysical, and biological processes of Antarctica are not significantly harmed by human activities." He explained further that because "the ecological structure of the Antarctic region is extremely vulnerable to disturbance," his colleagues endorsed CCAMLR, a novel treaty designed to protect Antarctica's whole marine "ecosystem."[101]

With apparently like-minded policymakers, natural scientists, and environmentalists as their main sources, journalists covering fisheries negotiations shifted tack and described Antarctica as a vulnerable environment rather than a robust and resource-rich frontier. One of the first articles on *Time* magazine's new "Environment" page reported in 1976 that "scientists fear that as the need for protein and minerals increases, peaceful exploration there may be followed by reckless exploitation." *U.S. News & World Report* offered a similarly eco-sensitive account, noting that the "U.S. position is that before resources are tapped, exhaustive studies to determine impacts on the fragile polar ecology should be made." Rounding out the field of major news magazines, *Newsweek* stated that "scientists and diplomats fear that even modest commercial exploitation could jeopardize both Antarctica's fragile natural environment and its delicate political ecology."[102] As national news media reoriented their editorial take on Antarctica, cracks in this new environmental consensus appeared in the United States as the CCAMLR took shape. US negotiators wanted to allow krill fishing right away, while environmental NGOs wanted to "get the research done first, and then evaluate rationally how much fishing can be done safely in Antarctica."[103] Their differences widened over the issue of Antarctic mineral development. Federal officials generally deemed such development inevitable, not to mention attractive to American prospecting companies, and they were once again prepared to allow such regulated activity. Environmentalists strenuously disagreed. They implored President Jimmy Carter in 1979 to endorse a regional "World Preserve" free of mineral development. They insisted that only a strict preserve could protect the "fragile terrestrial and extremely rich marine Antarctic ecosystems" and avoid "serious international political discord as a result of forcing confrontation on the sovereignty issue."[104]

The United States and its regional partners returned to that politically sticky issue in 1982 after signing the CCAMLR. The Consultative Parties expected an Antarctic oil rush sometime in the future, so they heeded SCAR's advice to establish regulations "before exploration interests develop."[105] As they earnestly embarked on closed-door negotiations, the Consultative Parties tried to head off criticism from several corners. They appealed to environmentalists by insisting that under any regime "protection of the unique Antarctic environment and of its dependent ecosystems should be a basic consideration."[106] They responded to critics of the Antarctic Treaty system by stressing that the "Treaty must be preserved in its entirety" and that its signatories "should continue to play an active and responsible role in dealing with the question of Antarctic mineral resources." They attempted to sideline the Non-Aligned leaders who coveted Antarctic resources as the Common Heritage of all nations by resolving that "the Consultative Parties, in dealing with the question of mineral resources in Antarctica, should not prejudice the interests of all mankind in Antarctica."[107]

The Consultative Parties had blunted objections to the Antarctic Treaty system and strongly endorsed nature conservation during negotiations for the CCAMLR, and they tried to repeat their diplomatic success with a minerals treaty. This time they faced mounting disapproval from critics in the United Nations and the environmental community. That community found its voice in 1977 when the first NGO dedicated to the region, the Antarctic and Southern Ocean Coalition (ASOC), agitated for a tightly regulated fishery and opposed mineral development in an Antarctic "World Park." By dint of ASOC's efforts, the International Union for the Conservation of Nature urged the Consultative Parties in 1981 to postpone a mineral regime "until such time as full consideration has been given to protecting the Antarctic environment from minerals activities." So too did the United Nations Environment Programme, which asked them to "give serious consideration to declaring Antarctica a World Park."[108] As the World Park caucus gained momentum, ASOC's James Barnes formed the Antarctica Project in 1982, an NGO to lobby Washington to protect Antarctica as "a key monitoring zone for global pollution, a scientific preserve for wide-ranging research of interest to all humans and a safe habitat for the largest population of wildlife on the planet."[109] Barnes passionately warned Congress that "multinational companies skilled in 'bending the rules' in order to achieve their acknowledged goal of making profits" would leave a trail of mine tailings and oil slicks that ruined the environment and with it the scientific utility of Antarctica.[110] While he emphasized the practical scientific value of a pristine Antarctic environment, Barnes made a still novel, romantic appeal when he called Antarctica "a symbol of hope to all people, a living reminder of the human ability to preserve its past, present and future, and to live in harmony with nature."[111]

US negotiators continued to express concern for the Antarctic environment, even though many officials in President Ronald Reagan's administration sparred with environmentalists like Barnes and eyed Antarctic resources as a boon for US oil companies. When US State Department representatives announced, in 1982, their hope that a treaty would "protect the full range of United States interests, including nondiscriminatory access for United States nationals and firms to engage in any permitted mineral resource activities," they maintained that "a cornerstone of the U.S. position in the negotiations is to allow mineral exploration and exploitation only if environmentally acceptable on the basis of adequate scientific knowledge."[112] Antarctic specialists in Washington understood that the region's scientific utility depended on environmental protection, as did the embattled legitimacy of the Antarctic Treaty system, so they continued to insist that this was a top concern. As a top negotiator explained again in 1984, the United States was determined that "resource activities, should they take place, take place in a fashion that is consistent with the protection of the Antarctic environment."[113]

Third World criticism of the Antarctic Treaty system also rebounded when mineral negotiations commenced in 1982. Malaysia's Prime Minister demanded that Antarctica and "all the unclaimed wealth of this earth must be regarded as the common heritage of all nations of this planet."[114] The United Nations General Assembly challenged the exclusive prerogatives of the Antarctic Treaty nations in 1985 and called on them to inform the UN Secretary General about their secret mineral discussions. The Consultative Parties rejected this political incursion and rebuffed the Assembly's request to "impose a moratorium on the negotiations to establish a mineral regime until such a time as all members of the international community can participate fully in such negotiations."[115] Although there had only been 14 Consultative Parties when this flak reappeared, the United States and its regional partners countered that the interests of the international community were in fact well served by the Antarctic Treaty. An American official pointed to United States and Soviet cooperation in Antarctica throughout the Cold War and the tolerance Great Britain and Argentina afforded one another there even as they fought a bloody war over the nearby Falkland Islands as proof that "the Antarctic Treaty represents one of the more successful examples of the practical implementation of the principles and purposes of the UN Charter."[116] A US diplomat introduced a new talking point in 1985 when he noted that the recent accession of Brazil, India, and China meant that Treaty signatories now represented two-thirds of the world's population, being made up of "developed countries and developing ones; Western democracies and Eastern-bloc nations; non-aligned countries, including a healthy representation of the so-called Third World."[117]

In staking equal claims to the continent's mineral riches, members of the UN General Assembly inadvertently advanced the cause of environmental protection in Antarctica. First, the Consultative Parties responded to their criticism by emphasizing their environmental credentials. They asserted that the Antarctic Treaty best preserved the peace and polar landscape and facilitated important scientific research on the global environment. Second, the United Nations provided environmental organizations with an international audience. ASOC and Greenpeace advised the General Assembly about Antarctic affairs and they used that body, according to one expert, "to get a better hearing for environmental/conservation concerns."[118] These NGOs no longer supported UN authority over Antarctica for fear this would set off territorial competition and damage the region's pristine environment. They still worked with the General Assembly to pressure the Consultative Parties to open up their private meetings and publish their resolutions. In an unprecedented move to blunt this coordinated challenge, the United States brought environmentalists, third-world critics, and journalists to Antarctica's remote Beardmore Glacier for a week-long workshop on the Antarctic Treaty system in 1985.[119] The United States had previously hosted "distinguished visitors" in

Antarctica to promote the USAP. This workshop represented a new effort to convert critics from the United States and around the world by showing them firsthand the regional peace and environmental research that might be lost if the Consultative Parties were not prepared to regulate an expected rush for Antarctic minerals.

Critical debate in the UN General Assembly did slacken, but a very different conversion occurred in 1989 when that body dropped its claims on Antarctic resources and urged "all members of the international community to support all efforts to ban prospecting and mining in and around Antarctica."[120] This political reversal may have been a last ditch attempt to derail the 1988 Convention on the Regulation of Antarctic Mineral Resource Activities (CRAMRA), which required several more Consultative Party signatures before ratification. The United Nations' reversal also reflected the fact that international concern about the Antarctic environment had recently and unexpected spiked. That concern erupted just months after the Beardmore workshop when a short article in the May 1985 journal *Nature* announced a substantial, 40 percent loss of high-altitude ozone over Antarctica.[121] Researchers had sampled stratospheric ozone levels in the region since the IGY. They had never observed such a precipitous drop. When NASA confirmed these findings and released satellite images of a gaping ozone hole over Antarctica, the alarming indication that industrial chemicals destroyed stratospheric ozone that blocked harmful solar rays focused international attention on the Antarctic environment (Figure 3.1).

This was not the first time Americans sounded the alarm over imperiled ozone. In 1973 the US chemists Mario Molino and F. Sherwood Rowland determined that chlorofluorocarbons (CFCs), a chemical widely used as refrigerants, solvents, and spray can propellants, could escape into the atmosphere and degrade high-altitude ozone. Although front-page headlines warned that skin cancer and eye cataracts would subsequently soar, Americans largely forgot this hot-button issue in 1978 when the United States banned CFCs in aerosol cans. The issue roared back to life when scientists pinned the Antarctic ozone hole on CFCs. The international community moved rapidly to phase out their production, but scientists warned that accumulated CFCs would continue to wreak havoc on the Antarctic ozone layer for many years to come.[122] Antarctica remained a remote and unforgiving landscape, but the psychic distance between it and the inhabited world collapsed as national media treated this casualty of industrial civilization as a harbinger of broader environmental peril. Antarctica once again clarified the acute vulnerability of

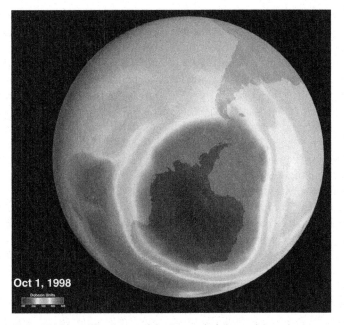

Figure 3.1   NASA satellite image of the "ozone hole" over Antarctica.

the whole planet, a lesson repeated in 1988 when scientists detected minor losses of stratospheric ozone over the Northern Hemisphere as well. As Democratic Senator of Tennessee Albert Gore rightly noted, that year registered a "sudden awakening of public concern about the global environmental crisis."[123] An American representative at international talks on CFC reduction gave voice to this public concern and acknowledged that "life as it has evolved on Earth is dependent on the existence of a thin shield." If people destroyed that ozone shield, he wrote, they could expect "millions of deaths from skin cancer, blindness from cataracts and injury to the immune system...and catastrophic damage to marine life and agriculture."[124] Aside from the potential disaster wrought by a nuclear war, the terrifying health and environmental costs triggered by ozone deterioration were rivaled only by those associated with another newfound crisis, the possibility that mounting "greenhouse" gases would rapidly alter the world's climate. As one scientist warned, global warming "raises the specter of considerable disruption to natural ecological systems, human agriculture and water supplies, threatens to raise sea levels or intensify hurricanes, and could cause unknown alterations to human and animal health."[125]

When Albert Gore campaigned for US President that year, he declared that the "depletion of the ozone layer, the greenhouse effect, and the global environmental crisis were the most important issues this country would have to face in the next decade and the next century."[126] His environmental message did not help him win that race, but it resonated in the national press. *National Geographic* devoted its final issue of 1988 to a worldwide environmental crisis and projected from its holographic cover the provocative question: "CAN MAN SAVE THIS FRAGILE EARTH?"[127] The editors of *Time* similarly consigned their year-end issue to what they perceived as an endangered planet. Instead of announcing its traditional "Man of the Year," the magazine dedicated the issue to the earth, its "Planet of the Year." *Time* editors justified their unconventional award by warning that "this wondrous globe has endured for some 4.5 billion years, but its future is clouded by man's reckless ways."[128]

Those reckless ways came off as the high road of progress just two decades earlier, when Antarctica stood out as a frontier for scientific scrutiny and high-tech conquest. Now news services and scientific journals emphasized the need to protect Antarctica as the best place on earth to monitor these environmental crises. A *Scientific American* cover story on the "Antarctic Ozone Hole," for instance, explained in January 1988 that Antarctic research was needed to determine if "the ultraviolet-absorbing layer is in jeopardy" since it had "revolutionized our knowledge of how ozone interacts with other gases and how the interactions are affected by meteorological conditions."[129] The springtime gap in the south polar ozone layer, *Business Week* wrote, had "catapulted Antarctica into a pivotal center for climate research."[130] Whereas a 1960 documentary cheerfully opined that "Antarctica is yielding to the march of machines and men," the 1991 IMAX film *Antarctica* saw the ozone hole as a warning that people had to stop that damaging march and use Antarctica to study "how we have damaged the planet."[131] In the first live television broadcast from the continent, ABC's "News Nightline" offered no timeworn references to an Antarctic frontier as it applauded scientists there for revealing that ozone depletion "is not nearly a local effect, it very much is a global effect."[132]

Like the ozone hole, the prospect of global warming continued to draw international attention to Antarctica. Scientists had assumed since the IGY that climate change would occur, albeit very slowly, causing Antarctica's ice sheets to grow or recede over thousands of years. However when a 90-mile chunk of ice broke away from an Antarctic ice shelf in 1988, *The New York Times* reported that many scientists then believed that "Earth's warming could already be speeding the flow of

Antarctica's ice."[133] Observers were quick to point out that this flow would accelerate ice melting. The consequent loss of Antarctic ice would raise global sea levels, inundating many of the most populated and productive coastal regions. A 1992 exhibit that toured US science museums argued this point closely and identified Antarctica as an imperiled stabilizer of the global climate. Since Antarctic ice "reflects sunlight back into space, preventing the planet from overheating," the exhibit explained, its recession due to global warming would reduce the continent's ability to cool the earth.[134] The children's magazine *National Geographic World* rendered this peril in age-appropriate terms, noting that the loss of Antarctica's ice would prevent it from acting "as a global air conditioner, cooling the planet and making it livable."[135]

As Americans turned their worried attention to the global environment, they learned that Antarctica was "particularly sensitive to the effects of pollutants on stratospheric ozone" and the place where "CO_2-induced temperature changes are likely to be largest."[136] As one scholar argued, because its "size and wider influence means that it cannot be ignored in a world which realises [*sic*] that global change affects all people," Antarctica should be protected "as a place to monitor global environmental pollution and atmospheric degradation."[137] Not surprisingly, environmentalists came to feel strongly that Antarctica was an indispensable site for studying ozone depletion and global warming. Greenpeace personnel were not yet focused on this issue when they brought their brand of direct action to Antarctica in 1987. They had then established the first nongovernmental and eco-friendly base there to promote "the region as a World Park, to minimize the effects of the human presence there and to prevent the exploitation of the area's mineral reserves."[138] They toured nearby Antarctic stations as self-appointed environmental watchdogs. Their well-publicized campaign jumpstarted the USAP's environmental cleanup and exposed France for dynamiting a new airstrip at its rocky coastal base right next to a sensitive penguin rookery.[139] As public debate about ozone depletion and global warming intensified, the organization linked its Antarctica campaign directly to these issues. A Greenpeace book explained in 1989 that Antarctica's "extreme vulnerability to change, highlighted by modern scientific research, makes it a sensitive monitor of the changes that humans are making to the biosphere."[140] It also advised Congress that strict environmental protection was needed to maintain this "unique laboratory to study such phenomena as stratospheric ozone depletion and the greenhouse effect," since in its "pristine state, Antarctica acts as a baseline against which science can monitor global pollution trends."[141] If the "Antarctic Treaty

Parties continue to hold open the possibility of extracting fossil fuels from Antarctica," a Greenpeace spokeswoman finally warned, "they will signal to the world that we are not yet prepared to take the steps necessary to reduce the threat of global warming."[142]

Many ranking politicians joined the fray. Most prominent among them, Senator Gore was saddled with the nickname "Ozoneman" by his Republican opponents for hammering on about the ozone hole and his related conviction that Antarctic science "represents our first line of defense against global environmental threats."[143] During a string of congressional hearings late in the decade, many of Gore's Democratic colleagues repeated his plea to protect Antarctica's threatened landscape as a critical platform for global environmental research. One called Antarctica "a global scientific laboratory of tremendous value" and a "center of research regarding ozone depletion and global warming."[144] Even President Reagan, whose wan regard for environmental causes was legendary and whose administration pressed for United States access to Antarctic minerals, acknowledged in 1988 that: "Through science we have seen that Antarctica is critical to the complex system of interacting processes that govern our environment."[145] As the NSF under Presidents Reagan and George H. W. Bush stepped up ozone and climate research in Antarctica, agency officials increasingly emphasized the contributions the USAP made to global environmental assessment.[146] The NSF's *Annual Reports* made these points for Washington policymakers, while its *Antarctic Journal of the United States* did so for the scientific community. The material it published for public consumption similarly observed that: "Dramatic springtime depletions of atmospheric ozone exemplify the region as a leading indicator of human-induced environmental change on a global scale," and explained that "with its icesheet—a ledger of past global environments—Antarctica, properly interpreted, can help tell us the planet's former conditions, its present status, and its likely future."[147] Even the State Department, which had spearheaded Antarctic mineral negotiations during the 1980s, acknowledged the critical importance of the south polar environment. As one official noted, the continent was the best place in the world to study "global environmental systems without the direct polluting effects of civilized society."[148]

The Antarctic nations faced relatively modest opposition in 1982 when they started mineral negotiations. Those negotiations ended with the CRAMRA, which would have allowed mining that conformed to existing Antarctic treaties as well as strict, if hard to enforce, environmental regulations. By the time the Convention opened for signing however, environmental NGOs had relentlessly targeted it as a threat to Antarctica and the global environment. As a leading scholar of Antarctic affairs then wrote, the "CRAMRA controversy, though centered primarily upon the future of mining in the world's last great wilderness area, has become part of a wider debate about the global environment."[149] International concerns about ozone depletion and global warming had thrust Antarctica to the center of this debate, but they had not yet broken the fraying consensus among Consultative Parties to implement CRAMRA. The final assault on that consensus began in March 1989, when Exxon's oil tanker *Valdez* ran aground in Alaska's Prince William Sound. The international media barely paid attention two months earlier when the Argentine transport *Bahia Paraiso* sunk and spilled more than 100,000 gallons of fuel near the Antarctic coast. The world took notice, however, when the Exxon *Valdez* clotted Alaskan waters with more than 11 million gallons of crude oil and vividly demonstrated how damaging the oil industry could be in pristine Arctic and Antarctic environments. *Sierra* magazine called Antarctica "Home of the cleanest water and the purest air on the planet," and it pointed to the *Valdez* and *Bahia Paraiso* spills as evidence that oil drilling would inevitably turn this "frontier for scientific research on the global environment" into a "Giant White Heap."[150] This was not just outlying bellows of an environmentalist periodical. The thoroughly mainstream *Time* magazine similarly warned that: "Once inaccessible and pristine, the white continent is now threatened by spreading pollution, budding tourism and the world's thirst for oil." Bracketed with pictures of Antarctic garbage dumps and oil slicks that stirred up fresh memories of the *Valdez* spill, it argued that only a prohibition on mineral development would protect the region's scientific utility and turn it into "the place where mankind finally learns to live in harmony with nature."[151]

CRAMRA's prospects evidently faded as even children's literature adopted a firm stand against mineral development in Antarctica. During the period of public alarm over *The Limits to Growth* in the late 1970s, the best-selling author Isaac Asimov brightly advised young readers that with Antarctica's abundant resources "the time may come when it will be very useful to humanity."[152] A decade later, children's books decried environmental decline in Antarctica, while *National Geographic World* warned its young readers that the southern continent "is the last true great wilderness left on the planet. If we destroy it, there is no way to predict earth's possible future."[153] The power of children's doe-eyed concern for Antarctica was evident in 1990 when international media paid close attention to French oceanographer Jacques Cousteau's expedition to the region. Lending his international celebrity to an eleventh-hour fight against CRAMRA, Cousteau brought six children with him, one from each of the world's inhabited continents, to make the case that the unblemished "Antarctic continent must be saved for future generations."[154] His high-profile voyage highlighted Antarctica's scientific utility. It also drew on the powerful symbolism of youthful innocence. Antarctica was the world's last uncorrupted corner that, like Cousteau's junior revue, warranted every measure of moral concern and paternal protection.

A powerful mix of practical conservationism and romantic environmentalism drew an uncomfortable light on CRAMRA, helping sink this hard-fought treaty. Environmentalists and their converts in the UN General Assembly insisted that only a strict ban on mineral development could preserve the incalculable utility and sublime effect of an unspoiled Antarctica. Realists in the US State Department remained unmoved as late as 1990, when they warned that the "permanence of such a ban could not be assured, and if it were broken, there would be no safeguards like those in the Antarctic minerals convention." They avowed that CRAMRA remained "the most balanced and environmentally sound framework that could be achieved to deal with possible interest in mineral resource activities in Antarctica."[155] But distant political currents finally doomed the mineral convention. After mounting environmental concern lifted Green Party candidates into office in Australia, that country changed course in November 1989 and refused to sign CRAMRA. So too did the governments of France and New Zealand. If valuable oil and minerals had been found in Antarctica by this time, their delegations may have embraced the convention so as to protect the polar environment and shore up the Antarctic Treaty. But those resources were still not apparent, allowing the Australian prime minister to high-mindedly reject "the clearly incorrect assumption— current in the 1970s—that mining in the Antarctic could be consistent with the preservation of the continent's fragile environment" and assert that the "most urgent and relevant action we can take is to ensure that this irreplaceable environment is never put to risk by mining."[156]

Under pressure from its Antarctic Treaty partners and the US Congress, President George H. W. Bush finally gave up on CRAMRA. In a rapid reversal, the United States and its Antarctic partners signed the 1991 Protocol on Environmental Protection to the Antarctic Treaty, establishing a moratorium on Antarctic mineral extraction designed to last at least 50 years.[157] The United States had pursued a mineral treaty for nearly a decade, stimulated by bright hopes for a Turnerian resource frontier and dark fears of territorial conflict triggered by an unregulated oil boom in Antarctica. But as an oil rush failed to materialize and Americans turned their attention from *The Limits to Growth* to global environmental crises, the United States changed course and followed the lead of its greener regional partners. President Bush did not betray his recent support for CRAMRA when he proudly announced that the provisions of the 1991 Protocol "advance basic U.S. goals of protecting the environment of Antarctica, preserving the unique opportunities Antarctica offers for scientific research of global significance, and maintaining Antarctica as a zone of peace."[158] These goals served national interests by healing divisions among the Consultative Parties and defusing outside criticism of their Treaty system. They also reflected a new paradigm of nationalist sentiment. The nation's preeminent program in Antarctica no longer evoked an American Century through the dream of planetary engineering or the prospect of a resource-rich safety valve for an overextended civilization. The USAP now demonstrated the nation's enlightened world leadership through its visionary commitment to protect Antarctica and use it to monitor the now fearsome threat of global environmental change.

The Once Terrible White Continent

In the decade after the 1957–58 IGY, proponents regarded the USAP as a minor pillar of the American Century, a pioneering venture to turn the last earthly frontier into a domesticated outpost of the Free World. Over the course of the 1970s and 1980s, they came to speak not about America's impending conquest of the southern continent but of its careful stewardship of Antarctica's fragile environment. As scientists revealed the region's sensitivity to distant industrial societies, officials heralded the USAP's soft-footed environmental research there rather than its heavy-handed conquest by heroic explorers. Whereas boosters and observers had prominently featured American men and their machines only 20 years before, by the late 1980s they reverentially treated what one book called *Antarctica: The Last Great Wilderness*, an area "relatively untouched by the ravages of man."[159] As ozone depletion and global warming "changed the world's attitude towards its environment," an expert on the region wrote in 1989, Americans construed Antarctica as the place where humankind could learn how to sustain its embattled planet.[160] They romantically depicted the "untamed island continent of Antarctica" as a moving symbol of unsullied nature, what one tourism company specializing in high-priced cruises to the continent called a "testament to the power and majesty of the natural world."[161]

The US Antarctic and space programs had attracted public interest early on, partly because they used some of America's most impressive technology. However, only space travel had in the rocket an example of what historian David Nye calls the technological sublime.[162] The spaceships that roared with volcanic intensity and streaked into the heavens were simply awe-inspiring and seemed the greatest measure ever of the nation's storied ingenuity and progress. Observers folded the USAP into the same narrative mold as space exploration by asserting that it extended American planetary engineering to the last terrestrial frontier. But they did not have a moving symbol like the rocket to evoke the grandeur of this visionary project. Instead, they dramatized the USAP by highlighting the brutal challenge of Antarctica and its terrifying, sublime environment. The sturdy men who braved its ripping winds and desperate cold, the NSF later admitted, conjured up "personal valor, danger, adventure, and of course, the hero" not for piloting its magnificent machines, but for contending with nature's most terrible arena.[163]

By the end of the 1980s, most pundits treated the naturally sublime Antarctic as an inspiring landscape and earth-nurturing environment rather than an alien, terrible realm. As *Popular Science* magazine explained, "far from being merely a white wasteland, a useless continent, Antarctica is vital to life on Earth."[164] That vital importance, one scholar hopefully predicted in 1987, would lead to "a possible re-orientation of values that has its genesis in Antarctica, a way in which man might come to regard the earth as a whole, politically, economically, and environmentally."[165] Only four years later the United States and its Treaty partners put their reoriented priorities into action by signing the Protocol on Environmental Protection to the Antarctic Treaty. Unlike environmental policies in the United States, which must accommodate competing interests and balance economic growth with nature protection, US policymakers followed an uncompromising tack in Antarctica. They could do so because Antarctica's unknown oil and mineral reserves remained beyond practical reach. They chose to do so to preserve the southern continent and protect the Antarctic Treaty system, which faced internal division and external scrutiny by Third World critics and by NGOs staunchly opposed to mineral development in the region. US officials accordingly treated Antarctica as a benighted symbol of the global environment and locus of America's new mission in the world. Rather than conquering the final terrestrial frontier with its mighty machinery and technocracy, the United States would demonstrate its power and enlightened leadership by protecting Antarctica and using it to tackle the world's most fearsome environmental crises.

The NSF embraced this emerging paradigm early on. Its 1988 report *Safety in Antarctica* admitted that Americans' cavalier actions at US Antarctic bases had exhibited an unenlightened "attitude of temporariness, of infinite environment." The USAP had mended its ways, the report avowed, and now exhibited an environmentally sensitive "mentality."[166] The NSF became a strong advocate for environmental protection, implemented a five-year "Safety, Environment, and Health" program in 1990 to cleanup its stations, and publicly aired its sordid history of environmental carelessness around its Antarctic bases (Figure 3.2).[167] Personnel at the US McMurdo station meticulously collected refuse for shipment to America and apparently honored the NGO that did so much to publicize Antarctic pollution by naming their recycling truck "Greenpiece."[168] Jacques Cousteau welcomed in 1990 what he hopefully saw as an improving global mentality, "the birth of a planetary consciousness" that recognized the environmental importance of Antarctica.[169] The strict measures adopted by the Consultative Parties in 1991 conformed to Cousteau's historical

Figure 3.2 With its diminutive scientist set against vast south polar ice, this 1990 National Science Foundation publication illustrates the emergent primacy of environmental study and preservation in Antarctica.

trajectory and fulfilled the desire of environmentalists everywhere "to know that somewhere on Earth there exists a whole continent that is almost pristine wilderness."[170] For Americans worried about the global environment, and especially those awakened to Henry David Thoreau's admonition that "in Wildness is the preservation of the world," Antarctica became a hallowed symbol of wilderness and of the hopeful fate of civilization. In "an increasingly polluted world," Greenpeace concisely explained, Antarctica represented "a place of pristine whiteness."[171]

This was not the profane whiteness once invoked by Herman Melville in his famous novel *Moby Dick*, the white tempests of nature that unsettle and sometimes destroy humankind.[172] Rather Antarctica's whiteness had become for many Americans a sentimental symbol of endangered wilderness, a planetary nest tragically befouled. The eruption of environmentalism and changes in US interests in Antarctica enabled this paradigm shift and spelled the demise of the trope of the Antarctic frontier and the slipping salience of American frontier nationality. Proponents of the United States space program clung to the frontier motif. But public observers saw a very different, albeit equally ennobling model of national character and purpose in Antarctica, where the United States endeavored to sustain the American Century through nature conservation rather than conquest. The USAP demonstrated the mettle of the nation not by domesticating the forbidding polar landscape but learning from and protecting the natural world in the now-sacred white continent.

Chapter 4

The Grip of the Space Frontier

When the editors of *Time* magazine announced an unconventional pick for its 1966 "Man of the Year"—not a single male but the generation of men and women 25 years old and under—they offered conventionally upbeat reasons for doing so. Brushing aside gathering clouds of war and student activism, *Time* suggested these American youth mostly avoided protest marches, accepted their nation's military action in Vietnam, and worked to positively transform their world. The editors predicted these young go-getters "will land on the moon, cure cancer and the common cold, lay out blight-proof, smog-free cities, enrich the undeveloped world and, no doubt, write finis to poverty and war."[1] Alas their rosy forecast did not come true. America's 12 lunar-landing astronauts turned out to be older than 25, and earthly afflictions of poverty and war endured. But the magazine's predictions echoed the still common conviction that an American Century of global peace and prosperity had begun. This new generation was ready to take charge and share the nation's good fortune and heal a divided world. Among the many endeavors that illustrated this promise, the US space program demonstrated quite dramatically that Americans young and old were able to outpace the Soviet Union and secure their benevolent leadership on earth and beyond. This according to many aerospace experts, public officials, and media pundits who agreed that the so-called space frontier constituted the ultimate phase of American development. By tapping a potent nationalist myth and suggesting that the conquest of outer space was analogous to the settlement of the New World and western frontiers, they promoted the US space program as a critical foundation of Free World security and prosperity. Although *Time* did not call its honorees a generation of celestial pioneers, its divinations fit neatly with everyday discourse linking the nation's frontier past with its spacefaring future.

Time pointedly took up the frontier trope two years later when it designated "America's moon pioneers," the three astronauts who first orbited that cosmic body on Christmas day 1968, as the "indisputable Men of the Year." Their "courage, grace, and cool efficiency" evoked the competent bravura of backwoods explorers lionized by nationalist chroniclers like Theodore Roosevelt. By asserting that the "newer world opened up by the Men of the Year will surely, in time, reach far beyond the moon," the magazine's editors also tapped a second strain of frontier mythology associated with historian Frederick Jackson Turner, who credited successive frontiers as the sustaining fount of American liberty and prosperity. Wonderstruck over a truly momentous event, the editors regarded the impending lunar landing not as a final goal but as the beginning of "a journey into man's future." They were sufficiently grounded to recognize that year's shocking political violence, urban unrest, and environmental degradation at home as well as war and nuclear proliferation abroad stood in stark contrast to that bright future and had made "it easy to question the wisdom of spending billions to escape the troubled planet." It turns out the editors were not clear-eyed enough. While they lauded the nation's pioneering future in space, these and further "upheavals and frustrations" cast shadows over the ensuing Apollo moon landings and shook the cultural and political foundations of the space frontier motif.[2]

Advocates had used that motif since the late 1950s to promote a budget-straining space program whose primary aim was to bolster America's security and prestige. This Cold War impetus foundered at the end of the 1960s as the United States won the race to the moon and pursued détente with the Soviet Union. The nation's simultaneous economic woes, triggered by its weakening trade position and enormous government spending, meant that the United States could not afford to carry out the ambitious program of human exploration envisioned by partisans of the space frontier. President Richard M. Nixon made this very clear when he declared "we must define new goals which make sense for the seventies" while realizing "that space expenditures must take their proper place within a rigorous system of national priorities."[3] Since those priorities did not include much touted space planes, orbital stations, moon bases, and piloted missions to Mars, many advocates of such plans had to follow Nixon's lead during that decade and stop summoning an unbounded space frontier. Rather than hailing an outward-looking nation of cosmic pioneers, they spoke prosaically of Americans as shareholders of a practical, earth-oriented space program.

Even without such constraints, these boosters may have dropped a space frontier motif whose inherited chauvinism and martial patriotism were ill-suited to a society grappling with racial and gender inequality, war in Vietnam, and the downsides of industrial progress. The trope of a forbidding Antarctic frontier poised for conquest by the United States withered owing to these very issues. The relatively few institutional players active in Antarctic affairs and journalists attentive to the region ultimately exchanged this nationalist paradigm of frontier conquest for one of environmental conservation and research. In this schema, the United States continued to demonstrate its righteous world leadership by protecting the now fragile southern continent and using it to study global environmental crises. The far greater number of star struck citizens, attentive journalists, and interest groups committed to a robust space program forestalled at least until the end of the Cold War any paradigm shift regarding outer space. Although cost-conscious officials laid off the space frontier motif, a gathering movement of citizen enthusiasts kept the dream of a spacefaring nation alive during the penurious 1970s. They argued that only America's wholesale colonization of the "high frontier" could revive its flagging world position and liberate humankind from its earthly troubles. During the next decade, NASA officials and like-minded industry spokesmen and media pundits promoted this dream once again. They embraced the rhetoric of Presidents Ronald Reagan and George H. W. Bush and exalted the space frontier, now pioneered by American men and women of all races, as a means of revitalizing the nation and securing its international leadership.

Downsized to Earth

The motif of the space frontier roared to life after the opening act of the space age. The Soviet Union's shocking October 1957 launch of Sputnik, the world's first satellite, called into question two pillars of US global leadership, its high-tech military superiority and unrivaled science and technology. The Soviet satellite tarnished America's reputation and unsettled the Cold War homefront, prompting political recrimination and front page anguish over the nation's humiliating loss in space. US officials expected to avoid such a crisis by beating the USSR into space with an America research satellite. Their failure plainly evident in the turbulent wake of Sputnik, federal powerbrokers hurried to make effective military use of earth orbit through secret Department of Defense rocket and satellite projects. They sought to further utilize outer space, boost national morale, and shore up

America's faltering reputation in 1958 by charging the new NASA with establishing US leadership in space and openly conducting civilian activities there for the high-minded "benefit of all mankind." As the Soviet Union scored a series of spectacular firsts in space over the next few years—the first animal, man, and woman in orbit; the first space walk; the first robotic impact of the moon—NASA's nearly billion dollar budget expanded several fold to finance the rockets, satellites, planetary probes, and manned lunar-landing missions that would, in addition to many other substantial benefits, prove beyond doubt that the United States was the foremost power on earth and in space. The epochal nature of the space age combined with the high drama of rocket launches and astronaut derring-do attracted enormous public interest in the late 1950s and 1960s. As America's aerospace complex spread across the country and counted hundreds of thousands of employees, NASA and its political, industrial, and professional partners worked to encourage that interest and turn it into lasting support for the US space program. They did so by itemizing the many mundane benefits of space operations. But they also found a more unified and visionary note better-suited to the nationalist impulses driving that program. These advocates hailed the space frontier for renewing America's vitalizing tradition of pioneering expansion, thereby allowing it to share its freedom and gathering prosperity with humankind.

The motif of the space frontier, born from science fiction and fantasy, became a sober selling point in the decade after Sputnik for the whole of the US civilian space program. Starry-eyed enthusiasts concentrated on manned spaceflight, owing to the sheer profundity of human travel beyond the earth. They also treated that program's complement of basic astronomical research, automated earth satellites, and planetary probes as coordinated measures for preparing the space frontier for human settlement. Boosters tied to the US space program clearly engaged in self-serving salesmanship. But the soaring motif of the space frontier also expressed their patriotic sensibilities and sincere embrace of an enduring nationalist myth at the height of its popularity. The strand of that myth associated with historian Frederick Jackson Turner held that America's exceptional prosperity sprung from its conquest of geographic frontiers. After the disorienting crisis of the Great Depression, the United States once again enjoyed several decades of economic growth that many people attributed to its new frontiers of science, engineering, and global trade. Advocates of the space frontier duly pointed out that path-breaking research, technological innovations, and tradable products and services spinning out of the US space program would drive this economic growth ever forward. Turner also argued that rugged frontiers had called forth Americans' natural inclinations toward individualism and democratic cooperation. Thus the vast space frontier held out the promise that liberty and democracy would prevail and indeed spread across the planet and beyond. Many public figures worried that Sputnik and subsequent Soviet space achievements portended a communist capture of this cosmic frontier, but Turner's story of liberalism's triumph suggested that, once committed, freedom-loving Americans would beat the Soviets and make full use of outer space. The martial undertones of Turner's account had been clearly drawn out by his peer and future president Theodore Roosevelt, whose take on the rejuvenating power of the frontier in the 1890s was very popular again in the late 1950s and 1960s. Roosevelt had glorified backwoodsmen of Anglo-Saxon and Teutonic descent as models for his devitalized contemporaries, for these pioneers had been the uniquely hardy and refined individualists capable of propelling their nation forward by civilizing savage lands. These were the stock heroes of Wild West fables, the cowboys and cavalry of Hollywood movies, radio serials, and television shows. Astronauts became their space-age counterparts. Cast in these media as pioneering men (all white) of steely nerves and steady hands, they indicated that Americans still had the individualist pluck and devotion to lofty national causes needed to surpass the hard-driving Soviets and conquer the space frontier.

The cultural currents that made that discourse sensible dramatically shifted at the end of the 1960s. The chauvinistic and martial strand of the frontier myth associated with Theodore Roosevelt took the hardest hit. Roosevelt straddled a thin line by celebrating America as a civic nation, an inclusive melting pot of multiple ethnic groups, while also treating it as a racial nation in which those Caucasian groups and select people of color could assimilate only by modeling themselves after their pioneering Teutonic betters. This fault between civic and racial nationalism in the space-age version of Roosevelt's frontier held as long as the all-white astronaut corps came off not simply as modern-day frontier heroes but also as color-blind envoys of the human race. The astronauts initially enjoyed this reputation, and leading African American voices seemed content for their peers to play less glamorous roles working for NASA on earth rather than in space. For instance, young civil rights activists in the Student Nonviolent Coordinating Committee then putting their lives on the line to integrate Mississippi's voter rolls may have dreamed of integrating the astronaut corps as well. But they simply appealed to NASA in 1964 to fairly review black job applicants at its new Mississippi rocket-testing center.[4] In the same vein, *Ebony* magazine had positively profiled "Negro" employees at NASA, such as a medical technician who looked after the astronauts, and a naval steward who broke new ground for his race while enduring the "grueling confinement" of a seven-day mock lunar mission.[5]

By the time lunar simulations gave way to real landings in 1969, African Americans rarely settled for such supporting roles. After an intense decade and a half appealing for racial equality under the law, civil rights activists called on NASA to recognize blacks' commensurate abilities by integrating its elite astronaut corps. As a *Christianity and Crisis* essayist wrote: "We have demonstrated our belief that white-skinned, crew-cut types can do just about anything to which they set their minds. Must we not demonstrate our belief that brown-skinned, black-skinned [people] have similar capabilities?"[6] *Ebony* gave voice to an increasingly jaded black community the next year by bluntly acknowledging discrimination in the aerospace industry and pointing out that many African Americans viewed "the first moon landing...cynically as one small step for 'The Man,' and probably a giant step in the wrong direction for mankind."[7] Skeptics of that giant misstep demanded that some of NASA's billions be used instead to solve earthly problems besotting racial minorities, particularly the urban blight and impoverishment *Time* magazine had presumed would disappear. Those who supported a downsized space agency demanded that it at least achieve greater minority representation among its workforce, especially after it drew bad press, in the words of the *Chicago Gazette*, for being "one of the most biased of all federal agencies." NASA faced that criticism in 1973 after it fired its "highest ranking black woman employee" for publicly denouncing that very bias. An agency official suggested that it hire a "highly skilled black public affairs type" to tell its side of the story and "help change the agency's negative image."[8] NASA prudently opted for the more substantive gesture of minority recruitment favored by aides to Presidents Gerald Ford and Jimmy Carter.[9] After the press "raked NASA over the coals for its equal opportunity shortcomings" and federal officers pushed it to recruit minorities, the space agency announced in 1978 that the first class of 35 astronauts training for the impending space shuttle included three black men.

These capable men represented a just change in the astronaut corps and an enlightened break from the racist, Rooseveltian vein of the space frontier motif. That vein rested on another chauvinistic assumption, namely that pioneering men, and men alone, had the strength and composure to advance the nation by conquering its forbidding frontiers. When the *Houston Post* reported in 1968 that the "American way of space travel has always been a man's job," it spoke of the reality of an all-male astronaut corps and the conventional belief that frail and nerveless women would soar beyond the earth only after those men had tamed the space frontier.[10] When 13 women passed the rigorous physical and psychological tests a NASA contractor devised in 1959 for the first team of astronauts, national newspapers joked about petticoated astronauts and a congressional representative paternalistically urged these "good ladies" in 1962 to "be patient."[11] Ten years later a female editorialist rued that "It's a shame that, in space, woman is still a joke." But that joke ebbed in the intervening years as the women's movement surged and feminists, who challenged prevailing norms of female domesticity on earth, dismissed the Apollo lunar program as the "Ultimate Phallic Journey" that implanted "Male Chauvinism on the Moon."[12] Under pressure of equal opportunity hiring statutes as well as women's rights advocates, including what one journalist called a "Legion of Angry Women" justifiably bent on breaking into the astronaut corps, NASA began recruiting women for its space crews and not just for its nursing and number-crunching teams.[13] When the agency picked six female astronaut-trainees in 1978 to join their three African American colleagues, the Rooseveltian trope of the space frontier had lost two of its most basic characteristics.

The racial and gendered profile of America's frontier past and space-faring future were traditionally intertwined with one more endangered aspect of Roosevelt's nationalist myth. The bully president had lauded white men, whose strenuous life of martial pursuit had immunized them against the enervating tendencies of modern society and prepared them for national leadership and competition with America's hard-driving adversaries. Colored by Roosevelt's gun-wielding version of frontier conquest and global economic expansion, the space frontier motif suffered further as Americans' martial patriotism wavered. If "the social function of Apollo was to sustain a pre-Vietnam dream of conquest," as historian Michael Smith contends, that dream of vanquishing illiberal enemies and taming unbroken nature lost its allure in the early 1970s.[14] By that time the country's ignoble failure to defeat communist forces in Vietnam signified its declining global hegemony and dampened Americans' frequent celebration of their nation's righteous war-making. Divided over the effectiveness and benevolence of US military endeavors, many Americans resisted further overseas campaigns and applauded the end of universal male conscription, which they had previously held up as a noble path to American manhood and national security. Desacralization of war and its bracing effects on young men was evident in popular culture. This was particularly true of the many Hollywood movies that recast traditionally heroic soldiers and western frontiersmen as perpetrators, and sometimes tragic victims, of US military aggression. Furthermore, although polls indicated that Americans continued to trust scientists and engineers even as their faith in other authorities tanked, the military mobilization of these professionals during the Vietnam War cast a similar shadow over them and fueled what the German missile engineer turned high-ranking NASA official Wernher von Braun called in 1971 a "climate of irrational hostility that seems to be growing in this country... regarding science and technology."[15] That hostility also emanated from corners of a growing environmental movement that regarded industrial-age science and technology as serious threats to the fragile natural world. Thus after years of war in Vietnam and of sniping criticism of science and technology as environmentally destructive and handmaidens of a brutal military, public celebration of a high-tech assault on the space frontier simply became anachronistic.

This left many space program boosters and media observers in a quandary. The former had used the frontier motif to sell their favored program, while they and like-minded observers embraced it as a sensible measure of a space program that carried the nation's venerable past into an even more promising future. Now that this motif no longer drew strength from Rooseveltian roots, its advocates were left to emphasize its Turnerian warrant. Historian Frederick Jackson Turner regarded America's settled frontiers as a lost source of socioeconomic progress. Liberated from the confines of thoroughly settled Old World societies, Turner declared, New World pioneers learned to cherish individual freedom and democratic cooperation while turning vacant frontiers into wealthy heartland. In the 1890s Turner rued for the future of American prosperity, not to mention liberty and democracy, now that the United States had been so widely settled that the US Census Bureau announced the closure of its last continental frontiers. While others recommended extraterritorial annexation or overseas trade as necessary replacements, Turner eyed science and technology as suitable alternatives to those defunct frontiers. Reliant on individual creativity and democratic collaboration, scientists and engineers were modern-day frontiersmen who would not only make exhausted hinterlands turn new profits but would also generate the knowledge and technological innovations sufficient for boundless economic growth. Turner's lab-coated pioneers never joined the ever popular western frontiersmen in the pantheon of public heroes. But their space-age descendants did. Although they often toiled in similarly workaday labs and drafting rooms, their celestial milieu was once again a spatial frontier, even more awesome and capacious than western lands. Working to open that vast territory, including what many assumed were habitable planets of Venus and Mars, aerospace scientists, engineers, and astronauts were akin to Turner's continental explorers who prepared vacant frontiers for subsequent waves of surveyors and settlers. The knowledge they gained and infrastructure they seemed ready to build in orbit, on the moon, and for piloted forays to other worlds would enable future Americans to move beyond earth and use the immense resources of space to engender freedom and prosperity for all humankind.

Although many aerospace experts and policy advisors had embraced this scheme and a wide spectrum of news and entertainment media had replayed it for nearly a decade, this space-age version of Turner's frontier thesis faced serious obstacles at the end of the 1960s as well. First was the sharp decline in what one historian called Americans' "grand expectations," the confident spirit that led *Time* magazine to predict in early 1967 the imminent conquest of nature, war, and poverty.[16] Mainstream voices had recently talked of engineering a more pacific and productive planet, even of controlling the weather around the world. Alarming reports of radioactive fallout, however, followed by front-page stories about pesticide contamination, oil spills, and polluted waterways silenced such blithe recitation about cost-free industrial progress and galvanized a growing movement to protect the natural environment. Far from ending war, the United States remained embroiled in a brutal Vietnamese conflict, and its heartland was seized by spasms of political violence and civil unrest. The urban riots that swept across the United States in the mid-1960s and reached a searing climax after the April 1968 assassination of Martin Luther King, Jr. exposed the terrible poverty and racial tension plaguing the nation's cities. With a flagging economy and a scientific and technological complex that had failed to preserve peace, prosperity, and the environment, many Americans understandably lost faith in the possibility of using that complex to conquer outer space.

Simultaneous with what one historian called this "waning of technocratic faith," the glamour of spaceflight dimmed as its novelty wore off.[17] Breathless predictions of America's mastery of a celestial frontier were far more common when barnstorming astronauts first made history by soaring into the sublime night sky. Their hypervelocity sojourns in the otherworldly firmament seemed an evolutionary leap, a mystical break from an earthly past that inspired visionary talk of illimitable prospects on the space frontier. NASA's robotic flybys of Venus in 1962 and Mars in 1965 were the first of many ensuing planetary missions that spoiled this romantic dream, for they showed that these desolate planets had neither inviting flora and fauna nor sustaining atmospheres many people long-suspected. Furthermore as astronauts' daring jaunts became more familiar, public observers began treating their orbital and lunar missions as routine orchestrations of a disaffecting technocracy, one incapable of solving Americans' terrestrial problems let alone casting people to all corners of an apparently barren solar system. As one enthusiast confessed, the "almost monotonous success of the [early manned space] flights...has evolved to near perfection with the Apollo flights since. Of course, that's a fact worth anybody's deep gratitude, but precision has a way of dehumanizing adventure."[18] The momentous occasion of America's first lunar landing during the July 1969 Apollo 11 mission temporarily rehumanized that adventure and stimulated front page discussion about America's spacefaring future. But even then many people shaken by their nation's profound trials turned their sights inward, away from an outward-pointing space frontier and back to earth. Rather than treating space exploration in Turnerian terms "as a liberating escape from the confinement to earth," they often assailed it as a costly diversion from earthly challenges or regarded it more positively, according to one literary scholar, "as a liberating return to fresh connection with earth."[19] This was an inward turn foreshadowed by *Time* magazine in 1968, which saw pictures of earth taken from lunar orbit as "urgent summons, in the words of Poet Archibald MacLeish, 'to see ourselves as riders on the earth together, brothers on that bright loveliness in the eternal cold- brothers who now know they are truly brothers.'"[20] President Nixon stuck to this message of terrestrial brotherhood during his congratulatory phone call to Apollo 11 astronauts on the moon. Rather than hailing their thrilling contact with the moon as humanity's first step toward settling that and other planetary bodies, he declared that their achievement "inspires us to redouble our efforts to bring peace and tranquility to earth."[21]

NASA Associate Administrator George Mueller resisted this cultural turn and used the momentous occasion of Apollo 11 to direct his nation's gaze back into space. Asking his fellow Americans "Will we press forward to explore other planets, or will we deny the opportunities of the future," Mueller answered with a space-age version of Frederick Jackson Turner's frontier thesis. Again, that thesis held that America's spirited conquest of continental frontiers had saved it from the economic torpor and social conflict associated with geographic confinement. Now that the US economy had soured and its environment and social fabric were under intense strain, Mueller gestured toward a space frontier as he warned his fellow citizens that if they forsook "the spirit of our forefathers then will man fall back from his destiny, the mighty surge of his achievement will be lost, and the confines of this planet will destroy him."[22] As NASA's budget continued its steady freefall, many program supporters appointed outer space a necessary safety valve from these planetary limits. Vice President Spiro Agnew did so obliquely when he endorsed a robust human spaceflight program in May 1969 and argued that "the nation should never turn inward, away from the opportunities and challenges of its most promising frontiers."[23] NASA Associate Administrator Homer Newell was more direct when he wittingly called space a "new frontier" for the human race that had "freed it from the chains that...have bound men to Earth" and promised "an essential stimulus to humanity's future development."[24] Calling this stimulus an "Absolute Necessity," a prominent aerospace engineer and tireless advocate of space exploration insisted that only this new frontier could avert an "apocalyptic...fate of a mankind endowed with cosmic powers and condemned to solitary confinement on one small planet."[25] Even a class of seventh graders who begged President Nixon to augment spending on lunar exploration did so because they saw the moon as a Turnerian safety valve, "a place to spread the exploding populous, to erect 'hot houses' to help feed the world."[26]

Whereas the two-stranded trope of the space frontier had reflected the buoyant optimism of a newly spacefaring nation, these advocates now used the largely Turnerian motif defensively to protect NASA's budget, which steeply declined after its 1965 peak. Attributing these cutbacks to heightened federal spending on social programs and the Vietnam War, NASA Public Affairs Officer Brian Duff determined in 1967 that the agency "must not allow itself to be shuffled into the back ranks because the nation's mass attention was temporarily diverted somewhere else." Fearing a loss of "attention and interest of the full body politic," he wanted to impress Americans that NASA's "work cannot be delayed without great loss to us and to all mankind."[27] Even as the space agency impressed Americans and a good deal of humankind two years later with the first of its six lunar landings, Duff's prescient concerns came to a head. Hundreds of millions of people watched the July 20, 1969 lunar touchdown live on television. A year later as many as 12 million Americans flocked to see the Apollo 11 capsule and samples of moon rocks as they toured America's 50 states. Millions more cheered the lunar astronauts as they made an around-the-world goodwill tour.[28] But through it all, national polls indicated weak public support for following the nearly $30 billion Apollo program with more costly and ambitious spaceflight plans. NASA's hopeful public affairs chief discerned "encouraging public support" from polls to maintain and even supplement NASA's substantial budget.[29] An independent researcher was likely closer to the mark when he determined that over the course of Apollo "the proportion of Americans who favored further government expenditures on space activities remained fairly steady at about two out of ten" while "those opposed gradually increased from about three out of ten to five out of ten."[30] This was in line with White House data, indicating that among Nixon's core constituency of well-to-do whites, "56% think the government should be spending less money on space exploration, and only 10% think the government should be spending more money."[31] Thus public support was anemic when President Lyndon Johnson's Science Advisory Committee recommended in 1967 that the United States gradually pursue a more substantial human spaceflight program, and even more so two years later when President Nixon's Space Task Group proposed that America's post-Apollo projects include launching a fleet of space planes, staffing an orbital station and lunar bases, and mounting a piloted expedition to Mars.[32]

Nixon rejected that proposal out of hand not simply due to flagging public support but also because the federal budget could not sustain such costly plans at a time when the Cold War impetus to race deeper into space had subsided. Surpassing the Soviet Union had been from the start a paramount justification for the US space program. Thus NASA Administrator Thomas Paine, who wanted to pursue those post-Apollo projects, predictably urged the president in February 1969 to give favorable "attention to the question of the future direction and pace of the nation's space program" since the "position in space of the United States relative to the U.S.S.R. is at stake."[33] But Paine's reasoning had become dated now that the United States was the undisputed leader in outer space, only months away from a lunar landing. In fact the previous administration had questioned that reasoning three years earlier, when USIA officials discerned that America had already claimed that mantle of leadership. Addressing "the question of whether to commit ourselves to still more ambitious programs—proceeding with manned exploration of the moon after the initial landing...[or] projecting man through space to the planets," State Department officials secretly determined in 1966 that "from the standpoint of our foreign policy interests, we see no compelling reasons for early, major commitments to such goals." They predicted further Soviet grandstanding in space but concluded that "interest abroad in the competition between the US and Soviet space programs, which is already diminishing somewhat, will lessen further" and become "focused more and more on the practical...applications of space programs."[34] President Nixon was even less interested in high-priced competition with the Soviet Union on earth and in space. He reduced costly Cold War commitments that taxed a sagging economy by scaling down troops in Vietnam and transferring security responsibilities to America's allies, whose growing purchases of US military hardware had the added bonus of reducing the nation's troubling trade deficit. He also initiated diplomatic relations with communist China and pursued détente with the Soviet Union, which slowed down a financially unsustainable weapons race and allowed the United States to choose more practical and cost-effective activities in outer space. Nixon's approval for a joint Apollo-Soyuz space mission, in which orbiting United States and Soviet capsules docked together in July 1975, epitomized détente and his effort to wind down an expensive cosmic rivalry.

That cost-cutting effort began shortly after his Space Task Force recommended an expanded human spaceflight program in September 1969. NASA's Thomas Paine lobbied hard for that program and Wernher von Braun insisted that it would open an economically stimulating "new frontier of space" that was "Vital to Man's Future."[35] Nixon bought von Braun's logic ten years earlier when the United States dominated the world economy and he urged Americans to strive as their forefathers had "from the earliest days of our history with the challenge of an unconquered wilderness and an apparently limitless frontier" so as to accelerate the science and technology on earth and in space "as necessary to human progress as it is to the security of free men."[36] The president now presided over an economy that could ill-afford such costly objectives and high-reaching rhetoric, so he put the brakes on this well-worn Turnerian play. Robust consumer spending and growing federal allocations for war and social programs, compounded by trade and currency imbalances, had battered the US dollar and triggered sharp inflation. President Nixon tried to right these imbalances by reducing federal spending and devaluing the dollar by freeing it from the gold standard in 1971, and he checked inflation that year with mandatory wage and price controls. Shortly after he lifted those controls in early 1973, the US economy stumbled again when the Organization of Petroleum Exporting Countries temporarily embargoed oil to many industrial countries, setting off a significant increase in fuel prices for much of the decade. In the face of persistent economic turmoil, many Americans reasonably worried that costly energy portended a dire future of dwindling natural resources as well as mounting pollution, a world to come in which "the limits to growth on this planet," according to the famous 1972 report *The Limits to Growth*, were fast approaching.[37] Thus as these challenges became high national priorities, federal patrons redirected a share of their downsized R&D allocations from Cold War projects to ones designed to stimulate America's economy, allay its social problems, monitor its environmental challenges, and augment its natural resources.

This was certainly the case with NASA, an R&D agency whose budget fell steadily after its 1965 high of $5.25 billion. An agency official later recalled that the president's budget advisors "were afraid that the enthusiasm of the country would create a runaway situation" in which NASA "would get everything they'd asked for."[38] But this was not to be. Public support for costly space exploration slackened, while Nixon's budget director downgraded human spaceflight in 1969 as a relatively low federal priority. Although the president loftily avowed that a "great nation must always be an exploring nation if it wishes to remain great," his March 1970 address on US space policy dwelt more on the "many critical problems here on this planet [that] make high priority demands on our attention and our resources." Nixon accordingly highlighted more affordable scientific research and focused on "practical application—turning the lessons we learn in space to the early benefit of life on earth."[39] White House advisors had already discussed applying NASA's technical and managerial expertise to the troubled "domestic scene," especially to facilitate large-scale redevelopment "programs necessary for the cities."[40] They even proposed "redefining the mission of NASA" to develop economically useful technologies, such as the ones capable of "desalting water" and thereby sustaining America's economically booming, arid Southwest.[41] Although these ideas were kicked around Washington throughout the decade, NASA retained its core aerospace mission albeit with practical-minded budgets that bottomed out at little more than $3 billion in Nixon's last year in office.[42]

By demanding practical results and imposing budgetary discipline, Nixon stifled further public talk about the space frontier.[43] *The Washington Post* regretted that the United States would not "press on rapidly in exploring the moon, sending men to Mars, building an orbiting space station, and beginning man's first attempts to probe deeper into outer space."[44] But the newspaper recognized that the "federal government faced . . . so many pressing needs in so many other areas" and it accepted the president's blueprint for NASA. Although his aerospace company stood to profit handsomely from such deep space endeavors, the chief executive of McDonnell Douglas admitted that "the National Space Program cannot be sustained at the cost levels of the past years." He suggested that it had to "be immediately related to such other national needs as the fight against pollution and poverty and the plight of the overcrowded cities."[45] Even NASA's Thomas Paine conceded in September 1970 that "with the lunar landing achieved, with America's concerns turning increasingly inward, and with competing budgetary demands by rapidly growing social programs, the current congressional mood was for diversified and practical space goals pursued at a moderate and economical pace." He therefore signed off on a NASA "program for the 1970s [that] reflect these desires, while doing everything possible within an austere budget to maintain our forward momentum in space."[46]

When Paine unveiled that program, he noted the president's strong support for ongoing space science in the 1970s. During that decade NASA launched X-ray satellites and solar probes, sent robotic observatories past Mercury and Venus, and most spectacularly of all placed a pair of Viking landers on Mars and launched two Voyager spacecraft on a grand tour of the solar system's outer planets. Paine put special emphasis on the "practical applications of space techniques" that Nixon clearly preferred.[47] Since industrial productivity and innovation were widely regarded as key to a healthy economy—one with lower inflation, better wages, and higher exports—NASA stepped up a program to transfer its path-breaking technologies to private US firms. As part of what one historian called a wider government effort "to foster the commercialization of civilian technology as a matter of broad public policy," the space agency sought out companies that might benefit from its knowledge and patented innovations and it made public the useful tools and processes derived from that R&D.[48] Rebutting skeptics who claimed that Americans had only Teflon pans and the orangey drink Tang for the billions they spent on space exploration, NASA's annual *Technology Utilization Program Report* highlighted its impressive contributions to the fields of medicine and physical therapy, fire safety and construction, energy and public transportation.[49] NASA's corporate partners also catalogued those industrial spinoffs, including the "new families of alloys and plastics, microminiaturized electronics, revolutionary fabrication techniques, [and] previously unattainable standards, tolerances and degrees of quality control" that have "grown out of space-related work and found their way into other areas of manufacturing."[50]

NASA officials also detailed their agency's practical value by appealing to Americans' growing concerns about environmental pollution and natural resource scarcity. At NASA's 1971 "Space for Mankind's Benefit" symposium, Wernher von Braun made the commonplace claim that astronauts' pictures of a colorful earth floating in the pitch-black void of space had raised people's ecological consciousness and even sparked the modern environmental movement.[51] As a like-minded editorialist explained, such photographs "gave man his first realistic and frightening stimulus to preserve a life-sustaining environment on earth and to begin an active fight to stem and reverse the tide of pollution that threatens…human survival."[52] Although modern environmentalism did not spring from these pictures, poignant new images of earth energized an already growing environmental movement, while NASA satellites provided natural scientists and concerned policymakers with timely data on worldwide environmental problems. A stream of press releases detailed how this useful data "can illuminate the obstacles to restoring productive harmony between man and nature."[53] Some of these obstacles were specific and local—a screw worm blight in Mexico and water depletion from the Florida Everglades—for which NASA's orbital eyes stood ready to help restore productive harmony. The obstacles of air and water contamination were more widespread. Since agency officials determined that "satellite systems play a valuable role in providing large scale overview" of such regional problems, NASA established a pollution-monitoring program in 1974.[54] By then environmental alarmists and sober scientists alike fretted over the grave prospect that global pollution was altering Earth's climate and depleting its stratospheric ozone layer. NASA stepped into the breach and used its weather- and earth-surveying satellites to monitor climate patterns, which many people blamed for worldwide droughts and crop failures. Worried that "a serious reduction in the ozone cover…could lead to an increase in the incidence of skin cancer, as well as changes in the average temperature of Earth's atmosphere," agency officials also mounted a "broad program of stratospheric research" that included satellite observation of high altitude ozone by 1979.[55] NASA's fittingly titled pamphlet *Improving Our Environment* summarily concluded that the agency was committed to help "manage the quality of our environment" by designing spacecraft "to observe and measure pollution locally, regionally, and globally."

That mission did not fit the Turnerian motif of the space frontier, which promised a starry escape from a confining, polluted planet. As long as continental frontiers existed, in Frederick Jackson Turner's schema, Americans had pell-mell exploited the land with little regard for the natural environment or resource sustainability, since they could abandon their exhausted fields, slash piles, and mine tailings and move to rich new ground. In one of its many environmental fact sheets, NASA's public affairs office rhetorically closed the door on a space-age version of Turner's thesis by admitting that this incautious dynamic was no longer prudent. "Conditioned by an expanding frontier," it quoted President Nixon, "we came only later to a recognition of how precious and how vulnerable our resources of land, water, and air really are." *Improving Our Environment* picked up this thread and asserted that satellites "help us better understand the forces that affect our environment," while "interplanetary exploration is helping to increase our understanding of our own planet." Rather than extolling NASA's planetary probes as frontier scouts, the pamphlet pointed out that data returned from Venus, Mars, and Jupiter "can be applied to increase knowledge about Earth's atmosphere and changes in Earth's weather and climate." It even dispensed with frontier references while touting the orbiting Skylab observatory, whose three crewed missions were the focus of NASA's human spaceflight program in 1973 and early 1974. "Skylab is another major step in remote environmental research," it explained, and "perhaps most important" among its many objectives was "to devise methods of gathering information about the earth surface that will enhance the well-being of mankind."

In this post-frontier stage of environmental accounting, NASA's new Environmental Resources Technology Satellites (ERTS) capably executed the dual task of tracking pollution and surveying natural resources. ERTS had "discovered air pollution sources in urban industrial areas...[and] photographed water pollution," the pamphlet explained, and had "recorded crops and croplands, provided data to refine a geologic map of petroleum provinces of Northern Alaska, and given indications of mineral deposits in other countries."[56] According to the agency's 1974 booklet *NASA and Energy*, these latter capabilities would help the United States deal with escalating energy prices and navigate the limits to growth posed by natural resource scarcity. Although it grimly confessed that oil and natural gas "will be exhausted" in the near future, it remained confident that the United States would avert ruin since ERTS could locate new energy deposits while NASA researchers worked on "economically attractive, renewable, and environmentally acceptable sources of fuel...as well as more efficient uses of fossil fuels until new energy sources become generally available."[57]

A post-frontier emphasis on cost-trimming practicality applied to human spaceflight as well. Although President Nixon declared in March 1970 that the United States should continue the Apollo moon program and "eventually send men to explore the planet Mars," he demonstrated his will "to reduce substantially the cost of space operations" by scrapping the last two scheduled lunar missions.[58] White House aides also considered canceling what turned out to be the last two missions to land on the moon, Apollo 16 and 17. The lunar program had been billed primarily as a scientific program, however, so the White House stayed that decision in part to avoid politically embarrassing criticism from scientists already facing sharp drops in federal financing.[59] Those aides also anticipated strong public support for these two missions and feared, most importantly, that such cuts would inflate aerospace unemployment in California, Texas, and Florida, states whose electoral votes were critical in the looming presidential election.[60] Among the more subtle and politically safe ways it signaled fiscal restraint, the administration edited the commemorative plaque for the final Apollo 17 mission. Since the White House wanted a phrase that did not commit the US "to additional moon shots," it vetoed NASA's proposed "We Will Come Again in Peace for All Mankind" for the noncommittal engraving, "May the Spirit of Peace in Which We Came Be Reflected in the Lives of All Mankind."[61] Set on the moon in December 1972, this plaque exemplified from afar what the journal *Nature* reported as a growing consensus for a more practical space program and more affordable "balance between the manned and unmanned aspects of the programme."[62]

Seeking that balance, the president shelved options for lunar bases and piloted missions to Mars and refused to match Soviet space stations with a new US orbital outpost. Defense planners had already deemed such an outpost unnecessary for national security since military reconnaissance was then conducted by satellite film photography and would soon be accomplished through real-time digital photography. Avoiding these gold-plated initiatives, Nixon encouraged NASA to design a cost-saving launch vehicle as a foundation for future human spaceflight. When a consultant determined that NASA's proposed fully reusable space plane would at best achieve minimal cost savings, Nixon approved a more modest if still revolutionary vehicle. In January 1972 he announced his support for a fleet of semi-reusable winged space shuttles that, by trashing their fuel tanks en route to orbit and gliding back to earth at the dizzying rate of 50 flights per year, would significantly reduce the cost of launching men and materiel into space. The estimated $5 billion shuttle fleet, Nixon averred, would enable future human exploration at the same time it was "a wise national investment" that would help "reorient our national space program so that it will have even greater domestic benefits."[63]

Since some editorialists and high-profile pols voiced opposition, including a 1972 Democratic presidential ticket that panned
the planned shuttle as "an enormous waste of money" that "will
deprive important social programs of much-needed revenue," the
New York Times acknowledged their dismissal of the launch vehicle
as "another grave distortion of national priorities...when so much
remains undone in meeting the needs of the cities, the environment
and the poor." Nevertheless, the paper endorsed the project as "a
major investment in the future" that will "alter the economics of space
activities and provide dividends that should continue for decades to
come."[64] So did CBS television's 1974 broadcast "Space: A Report
to the Stockholders," which itemized in businesslike terms how the
shuttle would meet national needs by cheaply servicing essential satellites and giving a listless economy the lift it needed to make new jobs
and improve the nation's cities and environment.[65] CBS likely drew
much of its information from the space agency, whose many publications and symposia identified how a shuttle-based program promised
"down-to-earth" benefits for Americans of "direct relevance to their
more immediate needs and pressing concerns." If a "continuously
expanding technological base [was] a prerequisite to the creation of
job opportunities," one symposium speaker explained, then the shuttle was a "cutting edge" path to "national technological leadership"
and "a favorable balance of trade in the face of increasing competition from nations with rapidly expanding industrial and technological capabilities."[66] So argued some of NASA's biggest cheerleaders. A
California senator praised the shuttle for its "tremendous foreign sales
potential" at a time when America's once mighty "trade position is
weakening," while an aerospace firm predicted a shuttle-based program would help "maintain a healthy economy, keep our world trade
position, solve our social ills, and build a better life."[67] For those who
wanted to know more specifically "What is the dollar return from the
space program, relative to the potential return from the same amount
of money invested in another sector of the American economy," NASA
officials answered in 1972 with a study that determined a seven-fold
return on each dollar the United States spent on aerospace R&D akin
to the shuttle. The news was even better three years later, when independent analysts calculated returns twice that rate and determined
that $100 million increases in NASA's annual budget would increase
US employment by 110,000, reduce inflation by 0.2 percent, and
stimulate economic growth.[68]

The budgetary constraints that engendered this unromantic func-
tionalism precluded more fantastic spaceflight schemes and compelled

NASA Administrator Thomas Paine, a missionary for such expensive projects, to resign in late 1970. "It had been widely rumored in Washington that the White House had wanted Paine out of NASA," historian Roger Launius explained, "because he was adamant in demanding increased funding for NASA while the administration wanted to hold the space agency's budget at a flat $3.2 billion a year." As one presidential advisor then wrote, "Paine may have had the ability" to oversee "a transition from rapid razzle dazzle growth and glamor to organizational maturity," but he "lacked the inclination—preferring to aim for continued growth." According to Launius, "Nixon wanted someone who was either in agreement with his goal of a smaller, less costly space program or a manager who would be more pliable."[69] White House advisors who vetted James C. Fletcher to be that manager must have been gratified when he told a home-town audience in Salt Lake City in 1972 that NASA was on track with a balanced and affordable program that served America's changing needs.[70] Those advisors were probably surprised, however, when this onetime corporate CEO and university president chafed at the utilitarian constraints that made his agency's budget, one presidential aide confessed, the most stringently controlled in the federal government.[71] It turned out that Fletcher believed that Americans' consuming attention to pollution and resource scarcity and their tanking faith in government, science, and technology had blinded them to the stupendous promise of the space frontier. In 1975 NASA's chief thus asserted, "in concentrating on the 'now' problems we are forced to ask questions about the future: are we losing sight of 'the dream?'"[72] He dreamed of a pioneering nation whose commonplace shuttle flights would allow it to build factories, utilities, and colonies in earth orbit, mine the moon and asteroids, and settle planetary bodies. Launius attributed this vision to Fletcher's religious compulsion to seek out extraterrestrial life and "protec[t] the earth through space operations."[73] But his theological impulse was not determinative, for a diverse range of citizen enthusiasts shared that same dream and kept it alive during the 1970s by arguing that a sidetracked United States could achieve greatness once again and lead the world out of its deepening morass by pioneering the space frontier.

Colonizing the High Frontier

James Fletcher was a politically sensible manager who lobbied for NASA by picking out its many practical returns. He was also a stargazer whose long-term outlook for the United States entailed an unbridled program of human exploration and exploitation of the solar system. Fletcher could not regularly preach, in the lingo of NASA officials, this "blue sky" sermon for fear of alienating NASA's thrifty patrons in Congress and the White House. But it became a mantra in the 1970s among citizen space enthusiasts who felt that Americans had turned away from a glorious celestial future by heeding critics of science and technology, cowering before prospects of resource exhaustion and environmental collapse, and penny-pinching NASA into largely earth-oriented applications. They believed that outer space was a frontier of vast economic potential in which pioneering Americans of all stripes could reclaim their nation's leadership and moral purpose. Since economic competitiveness, national leadership, and moral authority were guiding principles—or at least rhetorical keystones—for Ronald Reagan and his political allies, he gave a presidential imprimatur to the motif of the space frontier that allowed it to become conventional discourse once again in the 1980s.

The lay enthusiasts who kept Fletcher's dream alive during the 1970s engaged in a revitalization movement. They wanted to revive what they believed was a confounded nation and help it reclaim its providential destiny. They felt that the US space program had demonstrated that Americans had the talent and the science and technology to overcome the worldly constraints famously gauged in *The Limits to Growth*. Their faith was all the more urgent for it cut against a bleak vein of public opinion focused on the environmental and psychic costs of science and technology.[74] "Indifference to scientific achievement is the mood of the moment," *Time* magazine so lamented in 1973, due to an "increased awareness of the environmental ravages that seem to accompany technological advance" and to a "new mood of skepticism about the quantifying, objective methods of science." At the end of the decade, a presidential commission similarly regretted that the golden days when "we believed that we could do almost anything we set out to do" had passed owing to "a decline in public support for science and technology, closely related to expectations of material progress that seem difficult to satisfy."[75] Vexed by this apparent "climate of irrational hostility," space enthusiasts believed that science and technology could surmount earthly limits forecast not only by best-selling doomsayers but also by the *Global 2000 Report to the President of the U.S.* (1980), which grimly predicted: "If present trends continue, the world in 2000 will be more crowded, more polluted, less stable ecologically, and more vulnerable to disruption than the world we live in now." Even NASA's Voyager deep space probe carried this somber missive. In his recorded greeting to any extraterrestrial that improbably crossed paths with Voyager, President Jimmy Carter humbly suggested: "We are attempting to survive our time so we may live into yours. We hope someday, having solved the problems we face, to join a community of galactic civilizations."[76]

Public figures of all stripes contended that this first step, human survival, would come on what a more upbeat President Carter called "the gleaming wings of science," which alone could generate the new knowledge humanity needed to thrive on a competitive, resource-scarce, and fragile world.[77] But space enthusiasts emphatically reversed this calculus and stressed that survival and indeed a dazzling human future would unfold if the United States first pioneered the space frontier. Conservative political scientist Herman Kahn argued thus in his 1976 book *The Next 200 Years*. He dismissed as hogwash "rising concern about pollution and the possible exhaustion of natural resources" as well as commonplace predictions of "a much more disciplined and austere—even bleak—future for mankind." He rested his case on an "earth-centered" scenario in which American ingenuity engendered "a 'growth' world that leads not to disaster but to prosperity and plenty." However, Kahn speculated that a "space-bound" future entailing "large autonomous colonies in space involved in the processing of raw materials, the production of energy and the manufacture of durable goods" would be brighter still and "likely turn out to be closer to reality."[78] His space-bound retort to *The Limits to Growth* cribbed the substantial work of a gathering community of space colonization advocates. The foremost chronicler of this loose community of pro-space individuals and organizations suggested it did not exist only eight years earlier, when a "small, highly idealistic group called the Committee for the Future" extolled a new and timely vision of the space frontier as it lobbied Washington powerbrokers to let them take charge of surplus Apollo hardware.[79] Warning the White House and Congress that "earth is a closed system [in which] the only method of survival is total control," the group's spokeswoman asked them to steer away from this austere and statist future by allowing a citizens' lunar expedition to use that hardware to build a permanent base on the moon. These lunar homesteaders would take the first step in freeing Americans, "rich and poor alike, all who have a sense of hopelessness," from their planetary confines.[80]

This Turnerian rhetoric of transcending Earth's funereal limits found its greatest champion in Princeton University professor Gerard K. O'Neill. This noted particle physicist and once aspiring astronaut became an apostle of the space frontier after he asked his students in 1969: "Is a planetary surface the right place for an expanding technological civilization?" According to his telling, O'Neill and his students concluded the earth could not long sustain such a civilization. This did not mean he would stomach the belt-tightening measures later prescribed by *The Limits to Growth*. Instead his 1974 breakout article in *Physics Today* magazine, which finally answered his classroom query, sketched out a better future in which Americans transcended those limits by building earth-like habitats floating free in outer space. O'Neill developed these ideas further in his popular 1977 manifesto *The High Frontier: Human Colonies in Space*, a truly blue sky work that was sufficiently technical to be named best science book of the year by Phi Beta Kappa, the august academic honor society. He confidently asserted that Americans could quickly move beyond their burdened planet. The Apollo program demonstrated they had ample talent and technology to do so, while geological samples collected on the moon revealed that lunar regolith had most of the basic elements space settlers required. *The High Frontier* proposed the United States mine that soil and rock and use it to build vast colonies at the gravitationally stable "Libration" points near earth and the moon. These rotating cylindrical colonies would be far better than oft considered domed bases on the moon and Mars, whose occupants would endure tight quarters and limited gravity. O'Neill imagined that each of his spacious colonies would spin, thereby generating earth-standard gravity on their inside surfaces for 10,000 to as many as 100,000 residents. The rock-clad settlements would shield them from deadly solar radiation and afford plenty of room for greenery and spacious homes. Early spacefaring schemes rarely explained why people would cram together under glass and steel domes, assuming perhaps that this brand of celestial migration entailed an inviting urban modernism and a high-tech stage of frontier expansion. Gerard O'Neill was not so assuming, for his generational ideal of material progress centered on America's exploding suburbs rather than its troubled urban centers (Figure 4.1). His grassy colonies were more likely to excite fellow home-owning professionals. O'Neill regarded them as America's future pioneers, and he likely understood that they were an influential demographic needed to compel Washington to pursue his costly plans. Those plans were well worth the money, he averred, for the United States would revive an increasingly competitive economy to build these picturesque habitats. Each would be a "booming frontier settlement" that imported American "machines, tools, computers, and almost every other piece of complex equipment."[81] Like Turner's frontier entrepreneurs, these space pioneers would simultaneously enrich their national heartland by exporting mineral and energy resources back to their earthly homeland.

Figure 4.1   Life inside Gerard O'Neill's proposed human colonies in space.

This rosy economic blueprint revealed his utopian streak and accounted for the considerable backing O'Neill attracted. He admitted that industrial civilization had depleted Earth's natural resources and ravaged its environment, but believed with a decade-long crash program the United States could reap an untouched bounty of minerals on the moon and nearby asteroids. Residents of his celestial communities would pay their keep by building sun-powered factories to process those minerals for export to earth. In one grand stroke the planet's limits to growth would be breached and the global environment relieved as Americans harvested celestial resources and outsourced these most polluting industries to space. Those limits would be definitively overcome and earth returned to its Arcadian glory of ecological health when Americans shut down smoke-belching utilities and generated their electricity in space as well. Picking up aerospace engineer Peter Glaser's 1969 proposal to build satellite solar power stations (SSPSs) in geostationary orbit, O'Neill described how his high frontier workmen could operate giant utilities that turned a constant stream of intense solar energy into electricity, transferred to earth in microwave beams. He offered a bright future in which "the availability in the space habitats of high-paying jobs, of good living conditions, and of better opportunities," combined with "the availability in space of unlimited cheap energy, of abundant materials, and of efficient combinations of attractive living area with nearby industry" engendered an outmigration that made a "nonindustrial Earth with a population of perhaps one billion people...far more beautiful than it is now."[82]

Whether he realized it or not, Gerard O'Neill revived a winning combination of Frederick Jackson Turner's economy-boosting frontier and Theodore Roosevelt's morally rejuvenating hinterland. His scheme neatly mirrored Turner's thesis and its early space-age offspring by offering a grand account of the economic returns of the high frontier. O'Neill channeled Roosevelt's enthralling rhetoric to leaven a somewhat static, unromantic plan for pioneering suburbanism. He did so not by tapping the kinetic president's chauvinism or martial temper, for he made a point of inviting men and women of all races to enjoy the ample leisure afforded by his grassy colonies. Nor did he share Roosevelt's attention to individual heroics. O'Neill believed that his high-tech and management-intensive design would increase "individual freedom." But he was more a communitarian than rugged individualist, and he conceived his neighborly colonies as places where "bureaucracies become less important and direct human contact becomes more easy and effective."[83] What the community-minded O'Neill shared with the individualistic Roosevelt was a commanding sense that his frontier project had a vitalizing moral purpose. Theodore Roosevelt's aim was to rescue his nation from its spirit-draining excesses. Gerard O'Neill was not a martial proponent of the strenuous life. He wanted his fellow citizens to enjoy rich and fulfilling lives, and he sought to deliver industrial society from an earthly endgame. While Roosevelt hoped to fortify cerebral professionals with frontier pluck and strength, O'Neill cast these brainy suburbanites as America's pioneering vanguard ready to take to the stars and save the planet to boot.

Whereas public intellectual Lewis Mumford blasted O'Neill's "infantile fantasy" as a delusional manifestation of the technocratic impulse, many inspired observers fell for the high frontier. Such was the case with Stewart Brand. As the editor of *The Whole Earth Catalog*, a clearinghouse for environmentalists who espoused small-scale technologies and off-the-grid living, Brand was not an obvious cheerleader for O'Neill's grandiose scheme. Nevertheless Brand dismissed Mumford's fear of technological mischief and endorsed high frontier colonies as earth-healing mini-worlds that would engender therapeutic self-fulfillment and community building among modern civilization's alienated masses.[84] Aerospace engineer T. A. Heppenheimer was a more likely convert. Indeed his 1977 book *Colonies in Space* treated O'Neill as a tech-savvy prophet and echoed Stewart Brand by predicting that high frontier colonists would rediscover "a do-it-yourself approach" to life and prefer community-building venues such as "farmers' markets with open-air-stalls" to the colder economies-of-scale logic embodied in "glass-and-steel supermarkets."[85] While physicist J. Peter Vajik appreciated the prospect of "joyful, dynamic, and increasingly free" orbital societies as well, his straightforwardly titled book *Doomsday Has Been Cancelled* (1978) focused on O'Neill's sure-footed strategy to avoid a damning future of "dwindling energy and mineral resources, ever-widening starvation, accelerating environmental degradation, and more stringent social controls."[86] This hopeful outlook had attracted *The New York Times* in 1974 and explains why, in O'Neill's correct account, television "networks soon followed, newspaper and magazine reporters were not far behind, and a wave of public awareness and interest began to spread."[87] *National Geographic* featured his vision in its special edition celebrating America's bicentennial anniversary. Alongside the magazine's flag-waving stories of America's colonial founding, industrial development, and rise to global stature, its suggestively titled article "The Next Frontier?" featured free-floating space colonies as potentially the nation's next pioneering endeavor.[88] CBS's prime time TV news show *60 Minutes* offered a similarly bullish account in 1977 of how "some serious scientists are talking about whole colonies in space" populated by "hundreds of thousands of just plain folks looking to get away from an overcrowded earth, running short of energy, water and clean air."[89] When Florida's Disney World opened its theme park of technology futurism two years later, the Experimental Prototype Community of Tomorrow (EPCOT), one of its pavilions began offering simulated journeys to space colonies and SSPSs perched between the earth and moon.[90]

Such news and entertainment featured apparently serious proposals under consideration by scientific authorities. The learned journal of the New York Academy of Sciences ran an article in 1974 listing the many benefits of high frontier colonies, including their guarantee that "the human race will go on, even if there is a disaster on Earth, an environmental catastrophe or a nuclear holocaust."[91] Several years later the *Bulletin of the Atomic Scientists* presented a physicist's proposal to devote the impending Space Shuttle fleet to building a space colony and SSPSs, while *Technology Review* published a structural design of such a colony and a technical map for mining construction materials for space settlements from the moon and asteroids.[92]

Despite its politically prudent, practical, and earth-oriented canon, NASA cautiously supported O'Neill's ideas as well. When Wernher von Braun retired from the agency in 1974, NASA public affairs officers scrambled to find a new "far out spokesman" who could inspire an ardent constituency and articulate "blue sky" visions of "Large Space Stations, Scientific Bases on the Moon, and Satellite Solar Power Stations."[93] An agency official who wanted to build that constituency helped launch the National Space Institute the next year, for a time headed by the retired von Braun. According to *Science Digest*, this volunteer association aimed to stir "public and Congressional interest in all aspects of space development, including manned and unmanned space voyages, space industrialization and colonization."[94] In addition to such discrete constituency building, NASA officials sponsored a summer study on O'Neill's concept in 1975 that determined the "people of Earth have both the knowledge and the resources to colonize space," and they produced an animated film documenting the construction of high frontier colonies. Administrator James Fletcher was so taken with the concept that he excitedly sent a copy of *National Geographic's* bicentennial article "The Next Frontier?" to the White House, and his like-minded deputy opined in a keynote to Air Force alumni that their tricentennial speaker would likely address them from such a colony, such as that "proposed by Gerard O'Neill at the end of America's 200th year."[95]

Owing to mounting interest on Capitol Hill, the Congressional Office of Technology Assessment (OTA) looked closely at this issue in 1979 and determined that "space colonization, once a field of visionaries and science fiction writers, now attracts scientists, who advocate the mining of the moon and asteroids for raw materials."[96] In addition to scientists, the OTA could have identified several star struck members of Congress, attracted perhaps by the job-creating potential of space colonization, as well as a sizable bloc of citizens seduced by the revitalizing promise of the astral frontier. Clusters of hundreds and thousands of them joined many new space-oriented associations such as the L-5 Society. Named after the fifth Libration point, a site for O'Neill's proposed colonies, this organization formed in 1975 to promote stellar colonization as a means to "move most industry into space and return the earth to a garden planet."[97] The far larger Planetary Society, which counted more than 100,000 members, formed several years later to press Washington to fund basic astronomical research. Devoted to encouraging the exploration of the solar system, the organization also promoted human settlement of space, including the possible colonization of Mars. The enduring fan-base of the short-lived 1966–67 television series *Star Trek* was bigger still, for it supported regular conventions, successfully lobbied network television to broadcast reruns, and sent the Ford Administration "hundreds of thousands of letters" in 1976 "asking that the name 'Enterprise'" be given to NASA's first experimental shuttle. By acquiescing and naming the test-vehicle after that show's famous starship, the administration broke with its otherwise prosaic space policy. When NASA unveiled the shuttle Enterprise while a brass band played the soaring theme song to *Star Trek*, the agency signaled that the shuttle fleet would enable the United States, in the famous words of the TV show, "to boldly go where no man has gone before," beyond earth orbit and into "the final frontier." The Turnerian extrapolations of *Star Trek*, which cast the United States as the seed of a future federation of liberal planetary societies, exposed an enduring public enchantment with outer space. As President Carter's science advisor noted, so too did the 1977 Hollywood blockbuster *Star Wars*.[98] Although *Star Wars* indulged more in the Rooseveltian heroism of plucky freedom fighters, their planet-hopping struggles against a totalitarian empire excited an even broader public and indicated its fulsome romance with outer space and with America's martial frontier mythology.[99] The fewer but still considerable number of people who advocated space colonization expressed that romance in a more hard-headed way. They did not really believe that the United States could launch a federation of planets or defeat galactic tyranny. But they insisted that Americans could revitalize their nation and surmount worldly limits by pioneering the space frontier.

Proponents of the space frontier attracted national press, but they did not sway the White House, at least until Ronald Reagan took office. President Nixon often spoke in lofty terms about the US space program. His successor President Ford did so as well, and he praised it for calling forth "the best in the American character—sacrifice, ingenuity and our unrelenting spirit of adventure."[100] Still, both heads of state bowed to budgetary reality and kept a downsized NASA focused largely on space science and practical earth applications. So did President Carter, advised early on that while space exploration stirred the nation's soul and boosted its morale, high frontier proposals were merely "a technological solution to national problems and disappointments. It offers an ersatz frontier to replace the ones we have conquered and polluted," consultants to President-elect Carter explained, and "reassures us in international terms. It keeps us indisputably Number 1."[101] Like Presidents Nixon and Ford before him, Carter cautiously funded development of the shuttle fleet as the basis for future human spaceflight. But his administration's skepticism about large-scale space settlement and industrialization, underlined by the OTA's conclusion that they were "clouded by economic, political, and social uncertainties and by perplexing questions about the capability of technology to perform the tasks required," forced space agency officials to restrict their rhetorical support for such visionary undertakings. Thus agency officials had to backtrack on their support for O'Neill's work in 1978, when they defensively stated that NASA

financed studies "on a modest and continuing basis" to determine the
technical requirements of these still unattainable endeavors.[102] The
boldest measure they recommended was SSPSs. In a technical study
as well as two documentary films and multiple pamphlets, NASA suggested that the United States could reduce fossil fuel pollution and
achieve energy independence if it built these stations, each with solar
collectors spanning up to 20 square miles in geostationary orbit.[103]
Moreover, as one agency official confidentially noted in 1974, SSPSs
might "provide an ideal rationale for the shuttle program," a sentiment NASA's chief Robert Frosch frankly shared with Congress five
years later when he warned that "without some good solid program
commitments, such as solar power satellites,...the space program is
going to wither and die."[104]

Frosch may have been emboldened by a 1975 National Research
Council (NRC) report that determined the SSPS concept "appears
technically feasible." The concept provided a potential route to energy
independence that attracted serious interest from the head of the
short-lived federal Energy Research and Development Administration
as well as Democrat House Majority Leader Jim Wright of Texas. The
powerful Majority Leader asked the White House to give NASA $1.5
billion to develop an experimental power station.[105] President Carter
bucked this request and discounted warnings that a likely primary
challenger, California Governor Jerry Brown, might get a boost from
his endorsement of SSPSs. Carter duly announced in 1978 that it
was "too early to commit the Nation to such projects."[106] The president's wan regard turned cooler the next year when his budget planners determined "that $60 billion in federal funds, which would have
to be invested to achieve operational capability for an SPS system,
would be required at a time when the nation will need to make critical and massive capital investments in other energy technologies."[107]
If advocates of the space frontier held out hope their congressional
allies could wring SSPS funding out of an overstretched federal budget, that hope cratered in 1981 when the NRC revisited the issue and
concluded "that while a satellite power system is technically feasible"
its estimated $3 trillion price tag was "too costly to warrant its devel-
opment at this time."[108]

This cost-accounting prudence was a carryover from the 1970s,
when daunting economic challenges meant that public officials
could ill-afford to promote, let alone fund, such blue sky programs.
Devotees of the space frontier rejected this caution and urged their
contemporaries to engineer grand solutions on earth and beyond to
their nation's historic challenges rather than merely adapting to them.

A 1977 Rockwell International study on future space industrialization commissioned by NASA called this latter approach to national affairs a doleful attempt at "organizing scarcity," an unwarranted acceptance of material adversity by a mighty, frontier-minded country. The study contended that "the strength of a society is clearly revealed by the manner in which it manages and overcomes scarcity and want," and it urged the United States to reclaim its global leadership by aggressively exploiting "the space frontier."[109]

The Rockwell study was a portent of a political and cultural environment far more favorable for proponents of the space frontier. President Carter agreed that the US space program had enriched the nation and its global allies and would continue to do so in the impending age of the space shuttle. But the president took seriously what his chief domestic policy advisor called "An Era of Constraints," in which a flagging economy, partisan politics, and public cynicism limited the US government to judicious programs rather than financially risky, high-flown endeavors.[110] Ronald Reagan denied Carter a second term by rejecting this pinched logic and campaigning to return America to what he proclaimed its rightful greatness. He tapped an exhausted nation's nostalgia for a vital economy and global admiration, and promised to revive America's lofty purpose to advance human freedom and prosperity. Reagan extolled the rebounding US civilian and military spaceflight programs as means to these ends, and he lent the authority of his high office to the space frontier motif. Reagan's salutes to that motif allowed public officials and program supporters to pick it back up, turning the space frontier once again into a salient shorthand for national revival and enduring progress.

Pioneering the Space Frontier

Carter's actuarial realism distinguished him from Ronald Reagan, whose political coalition and winning 1980 presidential campaign rejected talk of national constraints. While Carter strained to manage foreign crises and a faltering economy, Reagan insisted that the United States could easily dispatch these myriad challenges. He exalted America as graced with a providential destiny and promised to reverse what he deemed self-induced degeneration and fruitless capitulation to the Soviet Union. Reagan rejected détente's logic of parity with the Soviet Union and wanted to reestablish America's superior military capabilities. The president did so by investing heavily in US military forces, beefing up strategic nuclear forces in Europe and Asia, and aggressively challenging communist-leaning parties in Latin America.

Turning back decades of public policy, Reagan attempted to stimulate the country's economy and make it more competitive by sharply cutting income and business taxes, deregulating industries, and circumscribing organized labor. Last he aimed to uplift a divided polity by encouraging patriotism and pushing conservative social policies. As one prominent historian explained, President Reagan consistently spoke about this diverse agenda in a "heroic idiom," one that framed his designs in familiar nationalist mythology.[111] Whether promoting his policy agendas or posing for pictures on horseback at his southern California ranch, the president favored above all the myth of the frontier as an enduring source of national prosperity and moral rejuvenation. His frontier allusions were often oblique, such as when he extolled Americans' storied self-reliance and entrepreneurial energy—like that exhibited by pioneering backwoodsmen—as cause for cutting the taxes, social welfare programs, and anti-discrimination measures he felt hobbled the country by dampening the entrepreneurial energy of hard-working people. Reagan did not cite America's frontier past when he defended the Vietnam War as a noble endeavor or when he pushed for a more muscular military posture. But he did tap that myth's martial chord, the Rooseveltian vein that celebrated military preparedness and action as righteous means to defend civilization from assault by savage enemies, be they Native Americans in the nineteenth century or globe-trotting communists during the Cold War. Reagan's frontier tributes were more explicit when he spoke of outer space. The president thus broke with his immediate predecessors and ardently embraced the motif of the space frontier.

That motif was not yet common currency when the Space Shuttle Columbia blasted into earth orbit in April 1981 on the fleet's long-awaited maiden flight. Thus when mission commander John Young "suggested that Columbia's journey brought man a step closer to the stars," *Newsweek* magazine speculated this was "not precisely what NASA's public-relations people wanted to hear" since they had "been promoting the shuttle in precisely the opposite way" throughout the 1970s. The magazine reminded readers that NASA had touted the shuttle "as the most practical, efficient, down-to-earth space vehicle ever designed, a 'space truck' whose mission is not exploration but the exploitation of the familiar region of nearby space."[112] But the recent induction of President Reagan augured a new moment when space-faring optimism, safeguarded during the 1970s by space colonization advocates, became common and credible once again. A *National Geographic* journalist anticipated Columbia's first flight by praising the shuttle fleet for "maintaining a frontier for us" without which the "country cannot grow."[113] Prolific science writer and editor Richard Lewis called forth the same Turnerian vision of frontier progress several years later when he insisted that the "Shuttle assumes evolutionary importance to the future of civilized societies" since it would open the solar system to industry and human settlement.[114] In his handsome coffee table book on the shuttle, a "long-time space enthusiast" channeled Frederick Jackson Turner when he said the spaceship would help Americans, "like the pioneers of the Old West, establish the initial settlements in space that will evolve into larger, more sophisticated facilities in the next century."[115]

Reagan rarely applied the frontier trope to his administration's fast-growing military space program. But that trope embodied the essence of his high-profile Strategic Defense Initiative (SDI), a grandiose plan to render ballistic missiles obsolete by building an orbital network of nuclear-weapons-destroying satellites. In line with Theodore Roosevelt's martial variety of frontier mythology, which held that a peaceful United States reluctantly picked up arms to defend itself in troubled hinterlands, Reagan introduced his SDI in March 1983 by reminding his audience: "The United States does not start fights. We will never be the aggressor. We maintain our strength in order to deter and defend against aggression—to preserve freedom and peace." America's continental frontiers had been the hallowed ground where the United States transcended confining boundaries and engendered that freedom and peace. Reagan hoped that the United States would continue to do so in space through his audacious plan to liberate "the human spirit" from the apocalyptic threat of nuclear war. If the world had survived for several decades under the terrible logic of nuclear deterrence and mutually assured destruction, the president preferred to rise above this logic in space and "begin to achieve our ultimate goal of eliminating the threat posed by strategic nuclear missiles." His critics called the plan "Star Wars," after the blockbuster Hollywood sci-fi movie, trashing it as a militarist fantasy whose technically improbable design would only raise the risk of nuclear proliferation and war. Their merits aside, those critics reinforced Reagan's frontier allusions with that name, for as scholar Tom Englehardt has argued, the *Star Wars* films rehabilitated the nationalist story of frontier conquest and once again made swashbuckling heroism in outer space a common feature of American popular culture.

President Reagan plainly leaned on the frontier motif in speeches about the US civilian space program. When he declared the fleet fully operational as the last "experimental" shuttle mission ended in June 1982, Reagan heralded that turning point as "the historical equivalent to the driving of the golden spike which completed the transcontinental railroad." If his reference was not crystal clear, it became so in 1984 when the president pronounced his hope "to build on Americans' pioneer spirit" and rapidly develop "our next frontier: space" by building an orbital station within a decade. Reagan proclaimed that the United States "has always been great when we dared to be great" and he heaped praise on the shuttles and station for allowing Americans to follow their "dreams to distant stars, living and working in space for peaceful, economic, and scientific gain."[116] NASA officials had lobbied the White House for a station for more than a decade. Now with presidential support, they energetically promoted it as a frontier outpost, a valuable scientific platform, gravity-free manufacturing center, and transit point for human voyages to the moon and planets.[117]

Although some quarters were deeply skeptical about an orbital outpost, Reagan proposed the space station at a time of heightened public interest in human spaceflight. Headlining that skepticism, *The New York Times* reported that the National Academy of Sciences "saw no scientific need for a manned space station for the next twenty years, and top military officials said they saw no unique military need for it." The newspaper further disclosed that intelligence agencies were "cool to the proposal, and the Office of Management and Budget vigorously opposed it."[118] Nevertheless, many thousands of Americans who belonged to more than a score of new associations devoted to US space exploration strongly supported the station. Millions more learned that the shuttle and orbital base were "the next logical step," in NASA's much publicized words, onto the space frontier. Those who traveled to Florida on vacation got this message at Disney's EPCOT Center, where several futuristic exhibits focused on outer space. The message was the same at the nearby Kennedy Space Center's Spaceport USA, located on what the state tourist board called "Florida's Space Coast," which was one of NASA's many visitor centers that celebrated the history and bright future of human spaceflight. Whereas only a handful of science centers existed a decade earlier, by the early 1980s many cities sported what one journalist called these "Amusement Parks of the Mind."[119] Part museum and part hands-on science education play place, these centers typically offered interactive displays about how the space shuttles and station would facilitate human exploration and settlement of the cosmos.

This was a prime topic for the Smithsonian Institution's National Air and Space Museum (NASM), one of the most trafficked museums in the world since opening in 1976. Every year millions of visitors saw stunning artifacts illustrating the impressive history of American aviation and astronautics, and they enjoyed entertaining accounts of the nation's future in outer space. The museum's 1985 IMAX film *The Dream is Alive* linked that future to Columbus's discovery of America and explained that with the shuttle and impending space station "some of our children will live in space, and their children may even be born there." That dream was alive, the movie explained, because "we now know how to live and work in space, we stand at a new threshold in the age of discovery." Among the many institutions vying to prepare young men for that dream was Admiral Farragut Academy, a military prep school whose mid-1980s advertising byline "Send your son to Mars" promised the training "they need to grab the best things life has to offer—perhaps even man's first steps on Mars."[120] Boys and girls could prepare to grab that spacefaring opportunity at one of several US Space Camps. Along with its sister programs in Florida and California, the flag-ship camp at NASA's Space & Rocket Center in Huntsville, Alabama has been "the granddaddy of all space-on-Earth experiences" since 1982 and given hundreds of thousands of participants a simulated "voyage into the unknown." An early book on the camp unmistakably titled *Your Future in Space* invited campers, "as part of our first real space-traveling generation," to prepare to "design spacecraft for interplanetary exploration, occupy space stations and help solve scientific mysteries that have baffled mankind for centuries." If this simulated "exploration into the possibilities and potentials that lie in [their] own future" was not recruitment enough, the 1986 movie *Space Camp* boosted camper applications by picturing on the big screen a rag tag group of campers who inadvertently rocketed to an orbital space station and piloted their shuttle back to earth in an emergency landing.[121] Several years later NASM's "Where Next Columbus?" exhibit offered a more studious portrayal of impending human planetary exploration. Most of the 300,000 people polled at exhibit computers felt that the "BEST reason for exploring space" was to "Increase knowledge and search for life." Their second most favored reason was to "Establish settlements or space colonies."[122]

The museum poll came after many years in which space-related entertainment, reporting, and educational programming encouraged Americans to consider outer space as their final frontier. This was true of The Young Astronauts Program, a privately funded program launched by the White House Office of Private Sector Initiatives in 1984. More than half a million Young Astronauts pledged "to get ready for the 21st century" by studying math and science delivered in curriculum oriented to space exploration.[123] President Reagan spelled out one of the program's central themes when he addressed its fledglings as the "generation that will move forward to harness the enormity of space" and, like their nation's pioneering predecessors, "expand the horizons of human freedom beyond the greatest dreams of our Founding Fathers."[124] Some of those Young Astronauts may have encountered similar material in college math and physical science courses. The 1985 academic conference "Space Colonization: Technology and the Liberal Arts" revealed that humanities and social science professors also used scenarios of interplanetary exploration and "space colonies to motivate students and faculty to look with fresh eyes on their particular academic material."[125] NASA had distributed this sort of educational material to grade school and college students for two decades. It did so again after the White House gave the space agency the green light in 1984 to hold a nation-wide competition to select a school teacher to be the "first private citizen passenger in the history of space flight." After NASA reviewed nearly 11,000 applicants, Vice President George H. W. Bush announced the winner as Christa McAuliffe, a New Hampshire high school teacher who compared herself to "the pioneer travelers of the Conestoga wagon days."[126] The lesson plans NASA prepared for McAuliffe's orbital broadcasts invoked the space frontier and asked students "to compare our future space settlers and pioneers to the early settlers and pioneers of America" and to consider if "migrations from Earth to Space Stations and other planets will be similar to the migrations from Europe at the turn of the century."[127]

Such expressions of frontier nationality reflected the rebounded spirit among parties of an aerospace complex then operating a fleet of space shuttles and designing an orbital station. That idiom projected their confidence, but it also gained force from troubling evidence that the shuttle had done little to secure the preeminence in space. While an idle US spaceflight program awaited development of the shuttle in the late 1970s, the Soviet Union dominated the field by using reliable old rockets and capsules to keep its cosmonauts regularly in earth orbit. The shuttle was far more sophisticated than Russian rockets and the proposed space station far more advanced than smaller Soviet capsules. But the USSR pushed forward into space with affordable, tried-and-true hardware and continued in the 1980s to clock many more person-hours in earth orbit, thereby amassing the extensive spaceflight experience needed for lunar missions and planetary expeditions. This was not supposed to happen. Shuttle boosters had insisted that the United States would quickly leap past its technologically antiquated rival and launch shuttles on a weekly basis. However when NASA called the now officially operation shuttles in 1983 NASA the "most reliable, flexible, and cost-effective launch system in the world," the fleet had failed to live up to this billing. It took more than five years to launch the 25 shuttle missions NASA officials once expected would easily fly in a mere six months. Furthermore, early game-changing budget estimates were never realized as shuttle flights cost, by many estimates, as much as ten times more than predicted. Spaceflight advocates who believed these poor showings were merely temporary could lean on Turner's version of the frontier motif and assert that these costly first forays, like those into the New World and western hinterlands, would soon allow the United States to find lucre on the high frontier.

The United States not only trailed its old communist adversary in space, it also faced pitched competition with rival providers of aerospace equipment and launch services. Successive presidents and members of Congress who supported the shuttle during years of development had been assured by NASA that the robust and affordable launch vehicle would maintain America's dominance of the global aerospace market by flying more than 50 missions per year. To achieve this vertiginous rate, the United States government wound down the expendable rocket industry and committed to using the space shuttle to lift nearly all its military satellites and commercial payloads. Federal subsidies helped the shuttle lift payloads for less than most expendable launch vehicles. NASA invited corporations and other countries to send their payloads aloft on the affordable space shuttle, which "demonstrated a remarkable suitability for delivering communications satellites to earth orbit." Its 1983 marketing brochure that touted the technological sophistication of an economy-boosting shuttle fleet failed to point out that higher than expected payload costs, stumbling launch schedules, and Pentagon priority for shuttle flights deterred many clients and allowed foreign competitors to take a sizable share of the global launch services market.[128] NASA Administrator James Beggs had similarly praised the shuttle-based space program in 1982 for sharpening the nation's "competitive edge based upon the continuing push on the cutting edge of technology." But he admitted that the program was then "competing with the Europeans day by day and fighting for new payloads to be taken into orbit." The European Space Agency's Ariane expendable launch vehicle that had become so successful since its 1979 debut that journalist David Osbourne called it "the world's only successful commercial rocket system." Since analysts expected global demand for aerospace services to grow sharply, Beggs understandably worried the United States would lose this important economic field to the Ariane, cheaper Soviet and Chinese rockets, and most ominously Japanese launch vehicles then in development. Expressing the timely fear that Japan might economically surpass the United States, Osbourne pointed out that Americans had "developed one new technology after another—from video-cassette recorders to machine tools to semi-conductors—only to watch the Japanese take the market from us." Just as the OTA had done, Osbourne warned that Japan and Europe were ready to steal America's leading position in the strategic and economically vital field of aerospace commerce as well.[129]

This flourishing commercial and human spaceflight rivalry flew in the face of the Reagan Administration's 1982 National Space Policy directing NASA to "maintain United States space leadership."[130] So President Reagan appointed members of the National Commission on Space in 1985 to fulfill Congress's charge "to formulate a bold agenda to carry America's civilian space enterprise into the 21st century." With members such as Thomas Paine and Gerard O'Neill, this blue-ribbon commission proposed a very bold agenda indeed. Their favored motif appeared throughout the commission's report *Pioneering the Space Frontier*, which summarily recommended that the United States "lead the exploration and development of the space frontier, advancing science, technology, and enterprise, and building institutions and systems that make accessible vast new resources and support human settlements beyond Earth orbit, from the highlands of the Moon to the plains of Mars." This agenda was rooted in the nation's basic fiber, for with "America's pioneer heritage, technological preeminence, and economic strength, it is fitting that we should lead the people of this planet into space." The study recapitulated O'Neill's space-age rendition of Frederick Jackson Turner's economy-boosting frontier and proclaimed that: "Now America can create new wealth on the space frontier to benefit the entire human community by combining the energy of the Sun with materials left in space during the formation of the Solar System."

The commission also conveyed Theodore Roosevelt's emphasis on the personal, liberating power of frontier exploration. It packed the report with graphic illustrations of bustling space settlements and people soaring freely across the starry firmament in personal rocket-suits. This technically inflated blueprint offered Americans an easy way to carry on their nation's frontier legacy and continue "to think, communicate, and live in freedom." Had *Pioneering the Space Frontier* limited itself to these blue sky scenarios, bookstores may have consigned it to their science fiction sections. But the report attracted serious consideration as a policy document that justified the space shuttles and impending station as parts of a coherent plan to build a permanent foothold in space. That foothold would lead to productive factories and utilities in earth orbit and ultimately to extraplanetary settlements.[131] In so doing it amortized the sky-high costs of the space shuttles and the mounting price tag for the station over many decades to come. It actuarially redefined these exorbitant spaceflight projects as affordable means to secure US leadership in the face of surging Cold War tension and global economic competition. For this reason, as well as its ill-timed debut, *Pioneering the Space Frontier* garnered front-page attention.

As the National Commission on Space prepared its final report, a launch-time disaster undermined its basic assumption that the United States could easily take to the space frontier. On January 28, 1986, the Space Shuttle Challenger exploded during liftoff and took the lives of its seven astronauts, one of whom was teacher-in-space Christa McAuliffe. As the paperback tribute *Heroes of the Challenger* noted, McAuliffe's addition to the most diverse American crew to date, one that included a Jewish woman and men of African and Asian descent, was meant to "make the space program more real and more attractive to those of us here on earth."[132] The millions of television viewers who watched the shuttle disintegrate on television learned that technical hazards and deadly risk were in fact more real than safe and citizen-ready spaceflight. The tragedy showed that the shuttle was not a futuristic Conestoga wagon reliably conveying pioneers onto a new frontier. It exposed a higher than expected probability of catastrophic accidents in complex technological systems, a lesson driven home three months later when the Soviet Union's Chernobyl nuclear power plant blew up and cast a radioactive plume across Europe. As aerospace engineers calculated a 1 in 78 chance of a similar shuttle accident, another round of debate whipped through Washington about national space policy and the fact the United States did not have backup expendable rockets adequate for its defense and commercial launching needs. The Challenger disaster made clear the strategic peril of relying so heavily on the shuttle, and the Reagan administration responded by resuming funding for alternative launch vehicles.

The disaster grounded the shuttle fleet and halted the human spaceflight program for more than two years. It was a profound shock and sobering turn for many crestfallen Americans. A 1987 NASA film captured their despair by conceding that the "space shuttle disaster brought to an end over three decades of success of what was perceived to be an almost superhuman ability on the part of the American production machine to set a goal, meet it and then with equal resolve exceed that goal and set another."[133] Never mind that the film overlooked the 1970s when such skepticism was more common, it recognized the general spirit of Reagan's presidential tenure when many Americans hailed the space shuttle, like their rebounded economy and emboldened military posture, as a sign of the nation's resurgence. This confessional film did not recognize what longtime critics as well as many supporters came to see as NASA's lax oversight and quixotic reliance on the costly and risky shuttle fleet. This was the stern message delivered by the bipartisan Presidential Commission on the Space Shuttle Challenger Accident headed by former Secretary of State William Rogers. The Rogers Commission determined that a failed pressure seal on one of the shuttle strap-on solid rocket boosters caused the accident, and it took NASA officials to task for disregarding contractors and agency personnel who raised red flags over this looming hazard. The commission tied this managerial negligence to "unrelenting pressure to meet the demands of an accelerated flight schedule," one that paled before an "early plan...of a mission a week" but was still frequent enough to show that the shuttle was indeed an affordable and reliable space ship.[134]

Despite this sensible criticism, post-Challenger polls revealed that a majority of Americans stood behind the shuttle and a leading US spaceflight program. According to one pollster, the "net effect of the Challenger accident was a strong shift of public sentiment in favor of the space program generally and the shuttle program in particular."[135] NASA public affairs officials were thus encouraged, and they regarded sympathetic editorials and congressional testimony as evidence that the agency's most influential observers remained keen on the shuttle-based, human spaceflight program.[136] Many spoke pointedly of the shuttle's critical importance to US leadership in outer space. Such was the case with the Rogers Commission, which recommended that NASA correct the space ship's flaws and use it alongside other launch vehicles "to serve the best interests of the nation in restoring the United States to its preeminent position in the world."[137] That finding was seconded by the Business-Higher Education Forum, a national association of corporate and university heads who saw the Challenger accident as evidence that "the U.S. lead in space is being threatened as the Soviet Union continues its ambitious space program and Europe and Japan move aggressively to harvest the bounty of space." Risk-analysis and cost-accounting cast cold light on the shuttle program, but the Business-Higher Education Forum pumped it up as an engine of the nation's high-tech economy, a key to industrial competitiveness and a first-rate university research complex. The Forum's heavyweights strongly backed the space shuttles and station and used language common after the accident to call on a vim and vigorous nation "to develop the boundless frontier of space."[138]

The bump in post-Challenger polls may have also derived from a powerful impulse to answer tragedy by rallying around the shuttle program. This impulse was central to Theodore Roosevelt's nationalist mythology, in which the frequent loss of life and limb turned backwoodsmen into national heroes, whose blood sacrifice gained exalted purpose only if Americans carried on their dangerous endeavors. President Reagan channeled this indomitable spirit within hours of the Challenger accident when he counseled shocked Americans that such loss was "part of the process of exploration and discovery." Advising them that the "future does not belong to the fainthearted," Reagan said the brave "Challenger crew was pulling us into the future and we'll continue to follow them." The unbowed president insisted: "We'll continue our quest in space" with "more shuttle flights and more shuttle crews and, yes, more volunteers, more civilians, more teachers in space."[139]

The next few years were touch and go for NASA. Its many critics saw the Challenger accident as proof that human spaceflight was a risky boondoggle that diverted precious funds from important military and civilian space projects and failed to maintain the nation's important aerospace leadership.[140] But the space agency enjoyed a bump in public support for its spaceflight program. Stalwart boosters in Congress such as Florida's Bill Nelson, who had flown aboard the space shuttle, interpreted favorable polls to mean that Americans "do not want a second-rate space program." Interviewees on a nationally broadcast documentary suggested that if the United States "lost its exploratory drive and was eclipsed by the vigorous exploits of others," namely Russia, Japan, and Europe, it would lose its global prominence and even "become a debtor nation in science by the end of the next decade."[141] As one astronaut declared on prime time television, the United States not only "needs to be at the forefront" in space, it was its "Manifest Destiny" to do so.[142]

Although the phrase "Manifest Destiny" smacked of discredited racial and martial elements of Roosevelt's frontier storyline, this astronaut echoed the common post-Challenger doctrine that the United States needed to cut out for the space frontier. This was a strategic imperative and an idealistic scenario reproduced in many books, including two Time-Life publications that tied old-time frontiersmen with future American pioneers poised to live in space colonies and travel on intergalactic spaceships.[143] The giant aerospace contractor General Dynamics added to these volumes in 1988 with a fictional issue of the *Planetary Explorer*, a make-believe journal of the real-world National Geographic Society turned "Planetary Exploration Society." Suggestively titled "The Emigrant Trail," this December 2038 issue looked back a half century and told the uplifting tale of how Americans rolled up their sleeves after the Challenger tragedy and reclaimed their leadership in space, sending men and women to the farthest reaches of the solar system and settling them on the moon and Mars.[144] The actual National Geographic Society had already given serious attention to this scenario and proclaimed the high likelihood that "Space Explorers Soon May Live in Self-Sustaining Biospheres." It cited as evidence the efforts of the private outfit Space Biospheres Ventures to build a self-regulating ecosystem in the Arizona desert. Advertising its sealed glass complex as a step toward "establishing permanent stations in space or on other planets," the group received a positive nod from NASA as well as many media outlets. This included National Geographic's children's magazine, which thought the project "may make it possible for humans to live on other planets."[145]

Officials at the Smithsonian Institution and Boston University's Center for Democracy took this possibility seriously and hosted a conference in 1987 to discuss the legal implications of space settlements and prepare a preliminary "Declaration of First Principles for the Governance of Outer Space Societies." According to the *Richmond Times-Dispatch*, many of the conferees came away "confident human settlement out there 'is desirable and inexorable'" and that it was then imperative to protect the legal rights of future space settlers. Supreme Court Justice William Brennan, Jr. left the conference convinced that "there will be space societies" in the near future needing the legal protections of "our constitutional heritage."[146] Even Capitol Hill reverberated with talk of impending space colonization when Democratic Representative George Brown of California presented "The Space Settlement Act of 1988." Challenging his fellow legislators "to prepare for a new phase in the space adventure," Brown proposed amending NASA's charter to acknowledge that "space is not only an arena for exploration and science, but also is an extension of our home, planet Earth."[147]

Buoyed after the Challenger accident by positive polls and this gush of public discussion about US spaceflight, NASA officials drove that discussion forward with learned reports that urged the United States to develop the space frontier. The 1987 study *Leadership and America's Future in Space*, produced by a team headed by the first American woman in space Sally Ride, admitted that America's "role as the leader of spacefaring nations came into serious question" after the Challenger accident. "Recognized leadership" remained critically important and "absolutely requires the expansion of human life beyond the Earth," the Ride Report proclaimed, "since human exploration is one of the most challenging and compelling displays of our spacefaring abilities." Citing a debt to *Pioneering the Space Frontier*, it identified the Space Shuttles and soon renamed "Space Station Freedom" as logical foundations of a program to conduct orbital surveys of the earth, launch planetary probes, and build a permanent lunar base. The Ride Report expressed "no doubt that exploring, prospecting, and settling Mars should be the ultimate objectives of human exploration," and it urged the space agency to examine further how the United States could implement such "an orderly expansion outward from Earth."[148] NASA responded with a new Office of Exploration to "coordinate agency activities that would expand the human presence beyond Earth, particularly to the Moon and Mars." Surveying the running debate about long-term spaceflight plans, *Astronomy* magazine welcomed the new office as a sign of "NASA's commitment to keeping that debate alive" and to engendering a national "goal of human expansion off the planet."[149] The Office of Exploration attempted to do so in a visually striking 1988 annual report, whose photo-realistic paintings of spaceships, lunar bases, and planetary excursions illustrated a proposed three-phase initiative to make the United States a great spacefaring nation. It confidently noted that the United States had already completed the first step of defining goals. Now it only had two phases to go: the design of propulsion and life-support systems during the 1990s; and launching lunar and planetary missions in the subsequent decade. NASA's *1989 Long-Range Program Plan* stood squarely behind this three-phase plan and detailed how the redesigned space shuttle and impending Space Station Freedom fulfilled the sober recommendations of the Rogers Commission and prepared the way for the visionary proposals of *Pioneering the Space Frontier*.[150]

This gathering string of space policy proposals reached a climax on July 20, 1989, when President George H. W. Bush commemorated the twentieth anniversary of America's first landing on the moon by calling on his fellow citizens "to establish the United States as the preeminent spacefaring nation" (Figure 4.2) Citing a familiar history linking "the voyages of Columbus to the Oregon Trail to the journey to the Moon," he summoned Americans to embrace that pioneering legacy and commit themselves "anew to a sustained program of manned exploration of the solar system and, yes, the permanent settlement of space." The president may have lacked what he called "the vision thing," the politically mobilizing ideas and ideology that his predecessor Ronald Reagan

Figure 4.2 President George H. W. Bush commemorates the twentieth anniversary of the first lunar landing by calling on America to become "the preeminent spacefaring nation."

famously possessed. But while Reagan mostly waxed about pioneering the space frontier, Bush endorsed an actual plan to make this happen, the so-called Space Exploration Initiative (SEI). According to this far-reaching initiative, the United States would use the space shuttle fleet to build the Space Station Freedom, followed by a permanent lunar station and "a journey into tomorrow, a journey to another planet: a manned mission to Mars."[151]

The SEI may have been unusually visionary for the pragmatic president, but it neatly complemented heightened public expectations about human spaceflight. The Republican Party platform of 1988 had reflected those expectations and foreshadowed candidate Bush's impending presidential action by declaring that: "A resurgent America, renewed economically and in spirit, must get on with its business of greatness." To do so, it had to "reestablish U.S. preeminence in space." More specifically, because "It's our nation's frontier, our manifest destiny," the United States "must commit to a manned flight to Mars around the year 2000 and to continue exploration of the moon."[152] This had been the prescription of the National Commission on Space and the proposed goals of NASA planning documents. It had been the nonpartisan advice of the heads of the National Academies of Sciences and of Engineering, who told President-elect Bush that "long-term, durable, and widely accepted goals for the nation in space are essential" and that those goals should include "a permanently manned space station [which] is needed to maintain a viable manned spaceflight capability for the United States."[153] The SEI also answered the heartfelt plea of the Republican co-chairmen of the Senate Air and Space Caucus, who urged Bush to "inspire and captivate a new generation of Americans, and citizens of the world as never before" by using "your leadership not only to set an ambitious space agenda for the years to come, but to start us on this course today."[154] President Bush seemed to do just that by proposing the SEI. His presidential commitment to an open-ended human spaceflight program indicated that spacefaring nationality and the associated motif of the space frontier had survived the hostile climate of the 1970s and once again enjoyed their claim as sensible expressions of America's character and destiny.

Conclusion

According to a NASA fact sheet, President George H. W. Bush's "exploration initiative was enthusiastically received by the space community."[155] That community had reason to be enthusiastic, for its fortunes had risen in recent years as military and civilian space budgets swelled. Popular news and entertainment as well as educators once again urged Americans to regard outer space as their great nation's destiny. Throughout the 1970s that community of aerospace professionals in government, industry, and academe was far less buoyant. They had largely abandoned the well-worn space frontier motif and the associated vein of spacefaring nationality evoked by President Bush in July 1989. Instead, they responded to America's changing culture, flagging economy, and weakened global stature and followed the lead of three presidents who preferred a less costly, more down-to-earth space program. Without the credibility afforded by aerospace authorities and government officials, the common motif of the space frontier nearly disappeared as the national conversation about space turned toward the practical. That conversation focused on an earth-oriented space program to boost a nation in crisis and help humankind adapt to its planetary limits. This sudden turnaround in public discourse reflected an equally sharp cultural turn among Americans whose gaze returned to earth and focused on humanity's vexing challenges of resource scarcity and environmental degradation.

As the nationalist paradigm based on scientific and technological prowess and on frontier conquest faltered, proponents of the USAP changed their tune about that national endeavor. They had treated the southern continent like outer space as a frontier for the American Century, a place where the United States demonstrated its benevolent world leadership by positively transforming humankind's domain, including its most forbidding frontiers at the ends of the earth and beyond. Doing so resonated with the optimistic spirit of the time and suited the Cold War purposes of the USAP. As that spirit waned and US interests in the region became more secure in the early 1970s, stakeholders in that program and unaffiliated observers struggled to redefine the USAP. Over the subsequent two decades, they emphasized environmental research and protection, which most Antarctic researchers deemed critical and American policymakers came to view as key to maintaining peace and national influence in the South Polar Region. By abandoning well-worn talk of conquering the Antarctic frontier, they reworked the nationalist paradigm of the American Century. At least in Antarctic affairs, the United States demonstrated its deserving world leadership by using its unrivaled power in Antarctica to study the global environment and protect the fragile southern ecosystem.

Whereas political and cultural forces compelled the national conversation about Antarctica to shift dramatically, the apparent turn in that conversation about outer space faced stiff resistance. National leaders in the 1970s demanded that a downsized space program serve people's immediate needs. But citizen enthusiasts who were raised on the astral promise of the early barnstorming era of human spaceflight argued that America should revive itself and save humankind from its earthly woes by pioneering the space frontier. Their stellar idealism, the antithesis of the utilitarianism of budget-wary leaders, helped seed a new synthesis for the space-age paradigm of frontier nationality. When Cold War tensions and global economic competition fueled this paradigm in the 1980s, public officials spoke of spaceflight simultaneously as a visionary project and a prudent investment for US security and economic competitiveness. As this message repeated across the echo chamber of public discourse—in industry circulars, professional journals and educational curricula, news and entertainment media—Americans seemed committed once more to pioneering the space frontier. The downsizing and earthly orientation of 1970s space policy appeared over, an aberrant moment of blinkered national pessimism definitively put to rest with President Bush's SEI.

The sad fate of the SEI revealed instead that disruptions in US space policy and more generally in American culture during the 1970s were not temporary. Americans had not momentarily and irrationally forgotten their nation's innate character and purpose, embodied in its lofty aim to secure an American Century of peace and prosperity on earth and beyond. Tectonic shifts in domestic and international affairs had forced them to reevaluate their national priorities and pretenses. The return of spacefaring nationality in the 1980s embodied the resurgent nationalist politics and culture of that decade pushed by national leaders and embraced by a polity uncomfortable with that reevaluation. But it did not last much beyond Bush's clarion call, for the Cold War came to an abrupt end and Americans learned that they did not have the money or technology, let alone the compelling purpose, to pioneer the space frontier.

Conclusion: The End of American Frontier Nationality?

Supporters welcomed President George H. W. Bush's Space Exploration Initiative (SEI) as a guarantee of national leadership long into the twenty-first century. A presidential advisory group accordingly hailed outer space as America's "most challenging frontier" and declared the SEI, by tapping "America's drive, ingenuity and technology—all those things that have made our nation the most successful society on Earth—will propel us toward a future of peace, strength, and prosperity."[1] This optimistic pairing of the nation's pioneer past and future capped a decade long revival of public enthusiasm for spaceflight and predictions that an American Century of global security and prosperity would continue under United States leadership.

Advocates applied that pairing to the US space and Antarctic programs in the late 1950s and 1960s. They tapped a popular national mythos to promote these expensive and strategically important initiatives. They also trumpeted these frontier endeavors as proof that America would forever outperform the Soviet Union and that its benevolent global leadership was aligned with its exceptional character and history. As media observers followed the lead of these national authorities, outer space and Antarctica became frontiers for the American Century, each a tabula rasa whose conquest would positively transform the human condition. When dramatic changes in domestic and international affairs disrupted this frontier discourse in the 1970s, program supporters and media observers channeled downsized public expectations by focusing on smaller scale benefits of these programs. As the strategic interests of the United States and its Antarctic partners cohered in the 1980s around environment research and protection, the US Antarctic Program once again featured as a civilizational project, now a critical effort to study and steward an endangered planet. Promotional buzz for the SEI illustrated that public discourse about spaceflight took a different tack during that decade as the associated motifs of the space frontier and the American Century revived and seemed poised to endure.

After all, the space shuttle fleet had returned to flight in 1988 and the space station initiative moved forward in the 1990s. It did so with Russia as an effective partner after the dissolution of the Soviet Union. Many Americans regarded the end of the Cold War as an epochal triumph and confirmation of the fact and righteousness of their nation's world leadership. For them, the promise of an American Century was evident too during the 1991 Gulf War, when American-led international forces quickly rolled back Iraq's invasion of neighboring Kuwait. As spectacular new genetic and information technologies emerged out of America's innovative economy and globalization accelerated, integrating it with the developing world and newly capitalist China, Russia, and Eastern Europe, there was ample reason to assume the United States would remain the indispensable leader on earth and beyond.

As it turned out, each of these factors played against the tropes of the space frontier and the American Century. Although President Bush saw the resumed shuttle flights as "American as Opening Day and as timeless as our history" and proof that "nothing lies beyond our reach," the fleet's constant delays and escalating costs subverted such lofty rhetoric.[2] The space agency's can-do reputation was further tarnished in 1990 when the newly orbited Hubble Space Telescope, advertised as America's eyes on the universe, turned out to have cataracts.[3] NASA's successful repairs made Hubble a national treasure that advanced science and gave humankind awesome images of the cosmos. However its fragility and whopping price tag highlighted once again the substantial costs and risks associated with spaceflight, reinforced later in the decade when the Russian space station Mir with Americans on board suffered so many glitches that the national press parodied it as a "Space Jalopy."[4] Mir may have been discounted as Russian kit, technically inferior to the impending International Space Station, but the ISS proved to be an exorbitant orbital outpost with an ambiguous mission and its own record of technical tribulations. Those problems and the tragic destruction of space shuttle *Columbia* on February 1, 2003 exposed once again the stark dangers associated with human spaceflight. By the time blockbuster movie *Gravity* (2013) vividly portrayed those dangers, cinematically depicting death and destruction in earth orbit, a two-decade cavalcade of US Mars rovers and international flybys and landings on other celestial bodies pointed to a future dominated by robotic rather than piloted exploration of cosmos.

Just as the diminishing luster of space shuttles and stations weakened the salience of the SEI, public attention turned away from spaceflight and toward other promising high-tech frontiers. For a few years after the Gulf War, that attention was fixed on advanced military hardware, particularly the guided missiles that seemed to knock enemy rockets out of the sky and precisely land US munitions on their intended targets. As these dazzling weapons systems failed to secure a Pax Americana, pacify so-called rogue states, or prevent attacks from terrorist organizations, two civilian fields of path-breaking American science and technology displaced public attention from aerospace R&D as they appeared ready to positively transform the world. America's highest profile big science program was then the Human Genome Project. This push to compile a complete database of the human genome attracted plenty of blue sky talk as boosters hyped the fast-growing biotechnology sector, attracting capital and potential markets to its profound and profitable applications for medicine, energy, and agriculture. Innovations in information technology were equally transformative and even more profitable, networking billions of people together through satellites and fiber optic cables and spawning the internet, arguably the most consequential new communication platform in centuries. Even though space enthusiasts predicted these innovations would help open up the celestial frontier by reducing the cost and risks associated with human spaceflight, industry advocates and public observers generally associated breakthrough genetic and information technologies with terrestrial progress. Their orientation reflected not only real on-the-ground economic opportunities and societal impacts but also the passing of starry-eyed futurism. That passing was evident in the late 1990s when Disneyland updated the area of the amusement park most closely tied to founder Walt Disney's streamlined visions of the future. Without an alternative, positive vision of the future that was culturally salient, Tomorrowland's archetypal rockets and space stations got a playful, retro facelift. Amusement park designers apparently believed millions of visitors would enjoy a nostalgic stroll through Walt Disney's astral futurism even if they no longer shared his belief that humanity's future resided in what he called "the new frontier, the frontier of interplanetary space."

As a prominent purveyor of technological futurism, Disney amusement parks had conveyed a promising storyline of American exploration and conquest of that new frontier for much of the Cold War. Disney's turn to nostalgia took place, not coincidentally, after the Cold War abruptly ended, undercutting the strategic importance of a robust human spaceflight program.[5] "With the end of the Cold War," the director of the Smithsonian National Air and Space Museum frankly noted in 1992, "the stimulus that gave rise to NASA is gone."[6] That

Cold War imperative to bolster the nation's defense, economy, and international prestige helped fuel aerospace investments and nationalist visions of America's space frontier. Without that stimulus, proponents of President George H. W. Bush's latest call to pioneer the space frontier failed to convince deficit-conscious legislators focused on a stalled economy to fund the SEI. As the Congressional Research Service then noted, "The bottom line is that until and unless the taxpayers of this country are willing to pay more in taxes or accept cuts in entitlement programs, there is little hope for the sizable increases in NASA's funding that would be required to pursue SEI in an expeditious manner."[7]

This congressional dismissal was notable for coming during the 1992 International Space Year (ISY), a bipartisan initiative conceived several years earlier to extol America's grand future in outer space alongside quincentennial celebrations of Columbus's discovery of the New World. Mundane fiscal concerns and mounting public debate about national history cast shadows over ISY festivities. As a *New York Times* journalist noted, Christopher Columbus had become a contested symbol. Whereas hagiographers praised him as an agent of New World progress, vocal critics called him a purveyor "of exploitation and imperialism."[8] Many of those critics promoted multicultural sensitivity to peoples long overlooked by standard historical narratives or drew inspiration from revisionist historians who recently challenged the celebrated myth of America's progressive frontiers by depicting what one scholar called its "unbroken past" of violent conquest in those lands.[9] With the Columbian and frontier myths in partial retreat, a *Washington Post* journalist critiqued the oft-repeated claim that America's space program confirmed "that we're pioneers, that we explore frontiers, that we use technology in that pursuit, and that we are a country with a special sense of our place in history."[10]

This fading public romance with the space frontier was symptomatic of a marked decline in visionary nationalism. The promise of the American Century, of a world made more peaceful and economically integrated under benevolent United States leadership, seemed very real at the end of the Cold War. Whether or not this moment constituted what one scholar called "the end of history," in which competition among liberal states replaced grand ideological conflicts, it was true that major communist powers had joined the international system they long opposed as capitalist economies.[11] By relieving the United States of the existential threat of a powerful ideological enemy, this triumph ironically undercut the nationalist paradigm that an indispensable America had an epoch-defining mission, on earth and in the space frontier, of leading humankind towards freedom and free markets.

In the absence of this paradigm, some people took Disneyland's path and waxed nostalgic about an earlier age of national power and purpose in outer space. Such was the case with the hit movies *Apollo 13* (1995) and *Space Cowboys* (2000), which glamorized a barnstorming era of spaceflight before NASA and the United States more generally became crimped by technological failure and bureaucratic enfeeblement. Many others urged the nation to embrace a new civilizational calling and save humankind from the sorry fate of the dinosaurs, whose demise was likely triggered by an asteroid impact. "Perhaps our space program will provide an 'insurance policy' for humans on Earth," one children's magazine speculated in 1992, so that "we can avoid our own extinction" by destroying incoming asteroids and comets.[12] This scenario took on urgency after data from US military satellites, declassified in 1996, revealed that nearly eight sizable extraterrestrial objects hit the earth each year between 1975 and 1992.[13] As many reporters, filmmakers, and government analysts noted, if any such object was large enough to kick up enough earthly debris to choke the atmosphere, it could doom humankind to a slow death by climate change.

Although a political divide over climate change intensified in the subsequent decade, many national leaders argued that the US space program, like the USAP, should squarely address the existential menace of global environmental degradation. Some of NASA's high-profile satellites monitored tropical deforestation, ozone depletion over Antarctica, ice loss in the Arctic, and the buildup of atmospheric carbon dioxide around the world. The space agency generally depicted its probes that visited Mars, Saturn, and Jupiter in the 1990s not as frontier scouts for piloted exploration—as they might have in the 1980s—but as astronomical data collectors that would help scientists better model the future of earth's atmosphere and environment. NASA's Mission to Planet Earth, an element of President Bush's SEI that survived congressional budgeting, supported that initiative with a multibillion dollar series of satellites designed to monitor earth's physical and biological systems. The Department of Defense, which incidentally needed to justify its post-Cold War space budgets, also recognized the strategic threats posed by environmental degradation. Thus it joined suit and pledged in 1995 to turn orbital swords into green plowshares by repurposing spy satellites "to gather clues about long-term global climatic change and ecological threats."[14] The changing mission of Biosphere II illustrates well this environmental turn. When a private venture built the giant glass structure in the Arizona desert in the late 1980s, it and NASA advertised the closed ecological system as a crucial test case for "establishing permanent stations in space or on other planets." But as the ersatz biosphere went sour, the venture determined there was no profit in extraterrestrial homesteading and handed the structure over to Columbia University. According to the *New York Times*, Columbia scientists used this mini-earth, subject to acid rain, greenhouse overheating, and biological obliteration, as a model for the "Atmospheric Nightmare" that could befall earth due to climate change.[15]

Thus at the end of the twentieth century it appeared that the cultural politics of US spaceflight might lean heavily toward environmental study and thereby realign with those of the country's Antarctic Program. In the latter case, as the National Research Council noted, the shift in America's priorities occurred as a "convergence of interests developed among scientific researchers, environmental groups, and the general public that look[ed] toward a responsible stewardship of," as well as research in, "the vast Antarctic land mass and its surrounding oceans."[16] The possibility that the nation might similarly devote itself to environmental study and stewardship from outer space never seemed as sure. As long as there had been a US space program, environmental research and observation had always featured among its myriad initiatives. They temporarily rose as a priority at the end of the century in public exchange due to a *divergence* of interests; national leaders, aerospace groups, and media observers no longer found common voice about how, or even if, spaceflight remained a measure of American power and its defining purpose in the world.

Starry-eyed visionaries continued to espouse the promise of the space frontier. But among these divergent interests and fractured voices of greater public authority, economic competitiveness and national security ebbed and flowed alongside environmental research as that defining purpose. The frontier motif popped briefly back into official discourse in January 2004, when President George W. Bush called on American to plot "a new course for America's space program." Like his father's SEI, the president's multidecade plan centered on a piloted program able "to carry astronauts beyond our orbit to other worlds."[17] Criticized by many as impractical drain on scarce resources needed to lead a global war on terror, Bush's proposal did not provoke widespread expressions of spacefaring nationalism. His ambitious call demonstrated, however, that as long as the United States sought or even reluctantly shouldered international leadership, the nation's frontier mythos might go dormant but will not likely disappear. Its familiar story of special character and purpose provides an inspirational identity needed for coherent national action. Thus as the country embarked on that global battle against terror, including a controversial and costly war once again with Iraq, the president reached for a familiar and uplifting narrative that cast the nation's aims as noble. His call to set out for other worlds clearly broadcast that the United States preferred spaceships over missiles and exploration over war and thus remained even in these troubling times a nation committed to a brighter future for all humankind. That lofty call did not survive the national exhaustion brought on by these wars and the deepest economic crisis since the Great Depression. But it will reappear if future presidents along with a convergence of interests see US spaceflight as an essential domestic and geopolitical priority, packaged best as a frontier project as grand as any America undertook during its pioneering past.

Notes

Introduction: Polar *Stars and* Stellar *Stripes*

1. *Let's Go to the Fair and Futurama,* New York World's Fair, 1964–65. Hagley Library, Pam 89.631; *Official Guide, New York World's Fair, 1964/1965* (NY: Time Inc., 1964), 77, 90, 204; Michael L Smith, "Making Time: Representations of Technology at the 1964 World's Fair," T. J. Jackson Lears and Richard Wrightman Fox eds., *The Power of Culture: Critical Essays in American History* (Chicago: University of Chicago Press, 1993), 223–244; Joseph Tirella, *Tomorrow-Land: The 1964–65 World's Fair and the Transformation of America* (Guilford, CT: Lyons Press, 2014).
2. James T. Patterson, *Grand Expectations: The United States, 1945–1974* (New York: Oxford University Press, 1996), 317.
3. Paul Dickson, *Sputnik: The Shock of the Century* (NY: Walker & Co., 2011, 2001).
4. *Let's Go to the Fair and Futurama,* 1964–65; *Official Guide, New York World's Fair, 1964/1965.*
5. Daniel T. Rodgers, "Exceptionalism," Anthony Molho and Gordon Wood eds., *Imagined Histories: American Historians Interpret the Past* (Princeton: Princeton University Press, 1998), 21–40.
6. Seymour Martin Lipset, *American Exceptionalism: A Double-Edged Sword* (NY: W.W. Norton, 1996); Robert Wiebe, *Who We Are: A History of Popular Nationalism* (Princeton: Princeton University Press, 2002); Andrew Delbanco, *The Real American Dream: A Meditation on Hope* (Cambridge: Harvard University Press, 1999), 45–80.
7. Henry Luce, "The American Century," reprinted in *Diplomatic History* v. 23, n. 2 (Spring 1999): 162, 166, 168.
8. Alan Brinkley, "The Concept of the American Century," R. Laurence Moore and Maurizio Vaudogna eds., *The American Century in Europe* (Ithaca: Cornell University Press, 2003), 11–12; Alan Brinkley, *The Publisher: Henry Luce and His American Century* (NY: Alfred Knopf, 2010); Stephen Whitfield, "The American Century of Henry R. Luce," Michael Kazin and Joseph McCartin eds., *Americanism: New Perspectives on the History of an Ideal* (Chapel Hill: University of North Carolina Press, 2006), 90–107.

9. David Chaney, *Scene-Setting Essays on Contemporary Cultural History* (NY: Routledge, 1994); T. J. Jackson Lears, "The Concept of Cultural Hegemony: Problems and Possibilities," *The American Historical Review*, June 1985, 567–593; Lee Artz and Bren Ortega Murphy, *Cultural Hegemony in the United States* (London: Sage Publications, 2000).

10. John Fousek, *To Lead the Free World: American Nationalism and the Cultural Roots of the Cold War* (Chapel Hill: University of North Carolina Press, 2000).

11. Robert Jervis, "Identity and the Cold War," Melvynn Leffler and Odd Arne Wested eds., *The Cambridge History of the Cold War, Vol. II, Crises and* Détente (Cambridge: Cambridge University Press, 2012), 22–43.

12. Luce, "The American Century," 171.

13. Brinkley, *The Publisher*, 7.

14. *Report of the Secretary of Agriculture* (Washington, DC: US Government Printing House, 1959), 60.

15. Wendy Wall, *Inventing the "American Way": The Politics of Consensus from the New Deal to the Civil Rights Movement* (NY: Oxford University Press, 2008), 163–200.

16. Kenneth Osgood, *Total Cold War: Eisenhower's Secret Propaganda Battle at Home and Abroad* (Lawrence: University of Kansas Press, 2006), 253–287.

17. Vannevar Bush, *Science—The Endless Frontier* (Washington, DC: US Government Printing Office, 1945), Cover letter.

18. David A. Hollinger, "Science as a Weapon in Kulturkampfe in the United States During and After World War II," *Isis*, 86 (1995), 442; Hollinger, "Free Enterprise and Free Inquiry: The Emergence of Laissez-Faire Communitarianism in the Ideology of Science in the United States," *New Literary History* 21 (1990): 897–910.

19. "Director's Statement," *National Science Foundation Seventh Annual Report for Fiscal Year Ended June 30, 1957* (Washington, DC: National Science Foundation, 1957), x.

20. Vannevar Bush, "The Essence of Security," December 5, 1949, Library of Congress (LOC), Papers of Vannevar Bush, Box 132, file Speech, MIT 12/5/49.

21. "Soviet Claiming Lead in Science," *The New York Times* (October 5, 1957), 2; "A Proposal for a 'Giant Leap'," *Life* (November 16, 1957), 53.

22. Testimony of Livingston T. Merchant, *Review of the Space Program*, House Committee on Science and Astronautics, 86th Cong., 2nd sess., January 20, 22, 25–29, February 1–5, 1960, 3.

23. Gretchen J. Van Dyke, "Sputnik: A Political Symbol and Tool in 1960 Campaign Politics," Roger Launius et al. eds., *Reconsidering Sputnik: Forty Years since the Soviet Satellite* (Australia: Harwood Academic, 2000), 396.

24. Statement by Lloyd V. Berkner, House Subcommittee on Territorial and Insular Affairs, June 10, 1960. LOC, Lloyd V. Berkner Papers, Box 21, Statement on Antarctica June 10, 1960 file.

25. Stuart W. Leslie, *The Cold War and American Science: The Military-Industrial-Academic Complex at MIT and Stanford* (NY: Columbia University Press, 1993).

26. Patricia Nelson Limerick, *Something in the Soil: Legacies and Reckonings in the New West* (NY: W. W. Norton, 2000), 74–94.

27. David Miller, *On Nationality* (Oxford: Clarendon Press, 1995); Ernest Gellner, *Nations and Nationalism* (NY: Oxford University Press, 1983); Eric Hobsbawm, *Nations and Nationalism Since 1870* (Cambridge: Cambridge University Press, 1990); Anthony D. Smith, *National Identity* (Reno: University of Nevada Press, 1991); John Hutchinson and Anthony D. Smith eds., *Nationalism* (Oxford: Oxford University Press, 1994); Geoff Eley and Ronald Grigor Suny eds., *Becoming National: A Reader* (Oxford: Oxford University Press, 1996).

28. Neil Smith, "The Lost Geography of the American Century," *Scottish Geographical Journal* v. 115, n. 1: 1–18.

29. Thomas Kuhn, *The Structure of Scientific Revolutions* (Chicago: University of Chicago Press, 1962).

30. *Pioneering the Space Frontier: The Report of the National Commission on Space* (NY: Bantam Books, May 1986).

31. "The End of the Space Age," *The Economist* (July 2, 2011), 7.

32. Bruce V. Lewenstein, "The Meaning of 'Public Understanding of Science' in the United States after World War II," *Public Understanding of Science* v. 1, n. 1 (January 1992): 45–68.

33. Eric Hobsbawn, "Nationalism in the Late Twentieth Century," Omar Dahbour and Micheline Ishay eds., *The Nationalism Reader* (Humanities Press, 1995), 370.

34. Dian Olson Belanger, *Deep Freeze: The United States, the International Geophysical Year, and the Origins of Antarctica's Age of Science* (Boulder: University Press of Colorado, 2006); Kenneth J. Bertrand, *Americans in Antarctica, 1775–1948* (NY: American Geographical Society, 1971); Stephen J. Pyne, *The Ice: A Journey to Antarctica* (Seattle: University of Washington Press, 1998); Christopher C. Joyner and Ethel R. Theis, *Eagle Over the Ice: The US in the Antarctic* (Hanover: University Press of New England, 1997); Aant Elzinga ed., *Changing Trends in Antarctic Research* (Dordrecht: Kluwer Academic, 1993); Edwin Mickleburgh, *Beyond the Frozen Sea: Visions of Antarctica* (NY: St. Martin's, 1987); G. E. Fogg, *A History of Antarctic Science* (Cambridge: Cambridge University Press, 1992); M. J. Peterson, *Managing the Frozen South: The Creation and Evolution of the Antarctic Treaty System* (Berkeley: University of California Press, 1988); Olav Schram Stokke and Davor Vidas eds., *Governing the Antarctic: The Effectiveness and Legitimacy of the Antarctic Treaty System* (Cambridge: Cambridge University Press,

1996); David Day, *Antarctica: A Biography* (NY: Oxford University Press, 2013); Paul Simpson- Housley, *Antarctica: Exploration, Perception, and Metaphor* (NY: Routledge, 1992); Christopher C. Joyner, *Governing the Frozen Commons: The Antarctic Regime and Environmental Protection* (Columbia: University of South Carolina Press, 1998); Kiernan Mulvaney, *At the Ends of the Earth: A History of the Polar Regions* (Washington: Island Press, 2001).

35. Among the sizable literature on the history and cultural history of US spaceflight: Howard McCurdy, *Space and the American Imagination* (Baltimore: The John Hopkins University Press, 2011); Martin Collins, *After Sputnik: The First Fifty Years of Space Flight* (NY: Smithsonian Books, 2007); Steven J. Dick ed., *NASA's First 50 Years: Historical Perspectives* (Washington, DC: NASA-SP4704, 2010); Steven J. Dick ed., *Remembering the Space Age* (Washington, DC: NASA-SP4703, 2008); Steven J. Dick and Roger D. Launius, *Societal Impact of Spaceflight* (Washington, DC: NASA-SP4801, 2007); Michael J. Neufeld ed., *Spacefarers: Images of Astronauts and Cosmonauts in the Heroic Era of Spaceflight* (Washington, DC: Smithsonian Institution Scholarly Press, 2013); Matthew H. Hersch, *Inventing the American Astronaut* (NY: Palgrave Macmillan, 2012); Margaret A. Weitekamp, *Right Stuff, Wrong Sex: America's First Women in Space Program* (Baltimore: Johns Hopkins University Press, 2004); James Kauffman, *Selling Outer Space: Kennedy, the Media, and Funding for Project Apollo, 1961–1963* (Tuscaloosa: University of Alabama Press, 1994); Mark E. Byrnes, *Politics and Space: Image Making by NASA* (Westport: Praeger, 1994); Walter A. McDougall, *The Heavens and the Earth: A Political History of the Space Age* (NY: Basic Books, 1985); William E. Burrows, *The New Ocean: The Story of the First Space Age* (NY: Random House, 1998); David Meerman Scott and Richard Jurek, *Marketing the Moon: The Selling of the Apollo Lunar Program* (Boston: MIT Press, 2014); Matthew D. Tribbe, *No Requiem for the Space Age: The Apollo Moon Landings and American Culture* (NY: Oxford University Press, 2014); W. Patrick McCray, *The Visioneers: How a Group of Elite Scientists Pursued Space Colonies, Nanotechnologies, and a Limitless Future* (Princeton: Princeton University Press, 2012); Linda T. Krug, *Presidential Perspectives on Space Exploration: Guiding Metaphors from Eisenhower to Bush* (Westport: Praeger, 1991); Roger D. Launius and Howard E. McCurdy eds., *Spaceflight and the Myth of Presidential Leadership* (Champaign: University of Illinois Press, 1997).

1   Rising to the Sputnik Challenge

1. Walter Sullivan, *Assault on the Unknown: The International Geophysical Year* (New York: McGraw-Hill, 1961), 1; Lloyd V. Berkner, "International Geophysical Year 1957–58," October 11, 1957. Library of Congress (LOC) Manuscript Division, Lloyd V. Berkner Papers, Box 12, File ICAF 11/Oct/57.

2. "A History of Science Policy in the United States, 1940–1945," Report Prepared for the Task Force on Science Policy, House Committee on Science and Technology, 99th Cong., 2nd sess., September 1986, 42.

3. Allan Needell, *Science, Cold War and the American State: Lloyd V. Berkner and the Balance of Professional Ideals* (Harwood Academic Pub., 2000).

4. Berkner memo for James R. Killian, "U.S. Policy and Action in Antarctica," February 14, 1959. LOC, Lloyd V. Berkner Papers, Box 17, file "Antarctica."

5. Glenn T. Seaborg, "Freedom and the Scientific Society: The Third Revolution," May 26, 1962. Wisconsin Historical Society Library Pamphlet Collection.

6. Henry Luce, "The American Century," reprinted in *Diplomatic History*, v. 23, n. 2 (Spring 1999).

7. Alan Brinkley, "The New Deal and the Idea of the State," Steve Fraser and Gary Gerstle eds., *The Rise and Fall of the New Deal Order, 1930–1980* (Princeton: Princeton University Press, 1986), 98.

8. Wyatt Wells, *American Capitalism, 1945–2000: Community and Change from Mass Production to the Information Society* (Chicago: Ivan R. Dee, 2003), 11.

9. *The First Report of the National Science Foundation, 1950–51* (Washington, DC: US Government Printing House, 1952), White House letter of transmittal.

10. Robert V. Bruce, *The Launching of Modern American Science, 1846–1876* (New York: Alfred A. Knopf, 1987); Roger L. Geiger, *To Advance Knowledge: The Growth of American Research Universities, 1900–1940* (New York: Oxford University Press, 1993).

11. Roger L. Geiger, *Research and Relevant Knowledge: American Research Universities since World War II* (New York: Oxford University Press, 1993).

12. Richard Rhodes, *The Making of the Atomic Bomb* (New York: Simon & Schuster, 1988); Thomas P. Hughes, *American Genesis: A Century of Invention and Technological Enthusiasm, 1870–1970* (New York: Viking, 1989), 381–421.

13. Vannevar Bush, *Science—The Endless Frontier* (Washington, DC: US Government Printing Office, 1945).

14. Daniel J. Kevles, "Principles and Politics in Federal R&D Policy, 1945–1990: An Appreciation of the Bush Report," Preface of Bush, *Science—The Endless Frontier* (Washington, DC: National Science Foundation, 1990), xii.

15. Derek J. de Solla Price, *Little Science, Big Science, and Beyond* (New York: Columbia University Press, 1986/1963); Peter Galison And Bruce Hevly eds., *Big Science: The Growth of Large-Scale Research* (Palo Alto: Stanford University Press, 1992); E. K. Hicks and W. van Rossum eds., *Policy Development and Big Science* (New York: North-Holland, 1991); and Robert W. Smith, "Large-Scale Scientific Enterprise," Stanley Kutler et al. eds., *Encyclopedia of The United*

States in the Twentieth Century (New York: Charles Scribner's Sons, 1996): 739–765.

16. Alvin Weinberg, *Reflections on Big Science* (Cambridge: MIT Press, 1967), 2.

17. Atomic Energy Act of 1946, Public Law 585, 79th Congress.

18. Stuart W. Leslie, *The Cold War and American Science: The Military-Industrial-Academic Complex at MIT and Stanford* (New York: Columbia University Press, 1993).

19. Melvyn P. Leffler, "American Grand Strategy from World War to Cold War, 1940–1950," Paul Kennedy and William Hitchcock eds., *From War to Peace: Altered Strategic Landscapes in the Twentieth Century* (New Haven: Yale University Press, 2000), 59; Melvyn Leffler, *For the Soul of Mankind: The United States, the Soviet Union, and the Cold War* (New York: Hill and Wang, 2008).

20. Truman Inaugural Address, January 20, 1949.

21. Thomas McCormick, *America's Half-Century: United States Foreign Policy in the Cold War* (Baltimore: Johns Hopkins University Press, 1995), 3.

22. Truman Inaugural Address, January 20, 1949.

23. John Krige, *American Hegemony and the Postwar Reconstruction of Science in Europe* (Cambridge: MIT Press, 2006).

24. Benjamin Fine, "Training of Foreign Students is Increasingly Important Function of U.S. College," *The New York Times* (June 27, 1954), E7.

25. John Fousek, *To Lead the Free World: American Nationalism & the Cultural Roots of the Cold War* (Chapel Hill: University of North Carolina Press, 2000).

26. *Historical Tables, Budget of the United States Government, Fiscal Year 2005* (Washington, DC: US Government Printing Office, 2004), 45–52.

27. Joseph Nye, Jr., *The Paradox of American Power: Why the World's Only Superpower Can't Go It Alone* (New York: Oxford University Press, 2002).

28. Luce, "The American Century," 169.

29. Clark A. Miller, "Scientific Internationalism in American Foreign Policy: The Case Study of Meteorology, 1947–1958," Clark Miller and Paul Edwards eds., *Changing the Atmosphere: Expert Knowledge and Environmental Governance* (Cambridge: MIT Press, 2001), 167–217.

30. Bruce L. R. Smith, *American Science Policy since World War II* (Washington, DC: The Brookings Institution, 1990), 66.

31. Ronald Fraser, *Once Round the Sun: The Story of the International Geophysical Year* (New York: Macmillan, 1961); Sydney Chapman, *IGY: Year of Discovery* (Ann Arbor: University of Michigan Press, 1968); J. Tuzo Wilson, *IGY: The Year of the New Moons* (New York: Knopf, 1961); Carl Eklund and Joan Beckman, *Antarctica: Polar Research and Discovery during the International Year* (New York: Holt, Rinehart and Winston, 1963).

32. Hugh Dryden, "The International Geophysical Year: Man's Most Ambitious Study of His Environment," *National Geographic* (February 1965), 285.

33. Sullivan, *Assault on the Unknown*, 4.

34. Dwight D. Eisenhower, "Remarks in Connection with the Opening of the International Geophysical Year," June 30, 1957.

35. Harold Bullis, "The Political Legacy of the International Geophysical Year," Prepared for the U.S. House Subcommittee on National Security Policy and Scientific Developments (Washington, DC, 1973), 13.

36. "Soviet Claiming Lead in Science," *New York Times* (October 5, 1957), 2.

37. Dwight D. Eisenhower Presidential Library, International Geophysical Year Digital Documents. National Science Foundation, "The International Geophysical Year," June 2, 1954, 4.

38. National Academy of Sciences, "International Geophysical Year: A Special Report," for the Senate Committee on Appropriations, 84th Cong., 2nd sess., May 28, 1956.

39. "Description of the Program," *New York Times* (June 30, 1955), A8.

40. Letter to Joseph M. Dodge, March 19, 1954. National Archives, RG 313, Records of the Commander, Admiral's files, 1954–56. Box 1, file "DOD Coordinating Committee on General Sciences."

41. Coordinating Committee on General Sciences, Report on US Program for the International Geophysical Year, May 19, 1954. National Archives, RG 313, Records of the Commander, Admiral's files, 1954–56. Box 1, file "DOD Coordinating Committee on General Sciences."

42. Rip Bulkeley, *The Sputniks Crisis and Early United States Space Policy: A Critique of the Historiography of Space* (Bloomington: Indiana University Press, 1991), 176.

43. CSAGI, *Bulletin d'Information*, No. 4 (London: IUGG Newsletter No. 9, 1955), 54–55.

44. D. W. H. Walton ed., *Antarctic Science* (Cambridge: Cambridge University Press, 1987), 33.

45. *The Secret Land*, Metro Goldwyn Mayer, 1948. National Archives Film Division, 200.310.

46. Rear Admiral Richard Byrd, USN, Retired, "All Eyes on the South Pole," *Reader's Digest* (December 1955), 117, 120.

47. JCS 1830/1, Report by the Joint Strategic Survey Committee to the Joint Chiefs of Staff on United States Antarctic Policy, January 1948, 9. National Archives, RG 218 Joint Chiefs of Staff Geographic File, 1948–50, Box 3, Antarctic Area Sec.1.

48. Senator Francis Case to Herter, July 1, 1957. National Archives, RG 59, General Records of the Department of State, 1955–59 Central Decimal File, Box 2773, file 702.022/4–157.

49. R. E. Byrd letter to the Secretary of Defense, "Importance of Operation Highjump II to Our National Security," August 29, 1949.

National Archives, RG 218 JCS Geographic File, 1948–50, Box 3, file Antarctic Area Sec. 2.

50. Confidential memo "What the Antarctic Is Worth in Relation to International Problems," June 2, 1947. National Archives, RG 218 JCS Geographic File, 1948–50, Box 3, Antarctic Area Sec. 1.

51. Secret "Memorandum for the Executive Secretary, National Security Council," August 29, 1949. National Archives, RG 218 JCS Geographic File, 1948–50, Box 3, file Antarctic Area Sec.2; Laurence Gould, *The Polar Regions in Their Relations to Human Affairs* (New York: The American Geographical Society, 1958), 15.

52. Secure Foreign Service Dispatch from the American Embassy of Moscow to the Department of State "Soviet Interest in Antarctica," February 7, 1955. National Archives, RG 59 Department of State, 1955–59 Central Decimal File, Box 2773, file 702.022/2–755; M. J. Peterson, *Managing the Frozen South: The Creation and Evolution of the Antarctic Treaty System* (Berkeley: University of California Press, 1988), 50–66.

53. Secret "Statement of Policy by the National Security Council on Antarctica," July 16, 1954. National Archives, RG 313, Records of the Naval Operating Forces, Formerly Classified Records, 1954–72, Box 1, file 1954.

54. F. M. Auburn, *Antarctic Law and Politics* (Bloomington: Indiana University Press, 1982), 89.These 12 countries included the seven claimant nations of Argentina, Australia, Chile, Great Britain, France, New Zealand, and Norway as well as Belgium, Japan, South Africa, the United States, and the USSR.

55. "U.S. Position on Antarctic Claims," April 3, 1957. National Archives, RG 218 Records of the US Joint Chiefs of Staff, Geographic File 1957, Box 1; John Foster Dulles telegram to Norman Armour, American Embassy in Paris, July 18, 1957. National Archives, RG 59 General Records of the Department of State, 1955–59 Central Decimal File, Box 2773, file 702.022/4–157.

56. F. M. Auburn, 91; Adrian Howkins, "Science, Environment, and Sovereignty: The International Year in the Antarctic Peninsula," Roger D. Launius et al. eds., *Globalizing Polar Science: Reconsidering the Social and Intellectual Implications of the International Polar and Geophysical Years* (New York: Palgrave Macmillan, 2010).

57. Representative Oren Harris letter to President Eisenhower, January 17, 1958. National Archives, RG 59 Department of State, 1955–59 Central Decimal File, Box 2773, file 702.022/1–658.

58. Confidential memo, "Antarctic Matters," October 21, 1955. National Archives, RG 59 Department of State, 1955–59 Central Decimal File, Box 2773, file 702.022/6–155.

59. Dian Olsen Belanger, *Deep Freeze: The United States, the International Year, and the Origins of Antarctica's Age of Science* (Boulder: University Press of Colorado, 2006), 36.

60. National Science Foundation and National Academy of Sciences Press Release, "Plans for Construction of Earth Satellite Vehicle Announced," July 29, 1955. Dwight D. Eisenhower Presidential Library, International Geophysical Year Digital Documents.

61. Statement by James C. Hagerty, Press Secretary to the President, July 29, 1955. Dwight D. Eisenhower Presidential Library, International Geophysical Year Digital Documents.

62. Sullivan, *Assault on the Unknown*, 163.

63. R. Cargill Hall, "Origins of U.S. Space Policy: Eisenhower, Open Skies, and Freedom of Space," John M. Logsdon et al. eds., *Exploring the Unknown: Selected Documents in the History of the U.S. Civil Space Program, Vol. I: Organizing for Exploration* (Washington, DC: NASA, 1995), 215–16.

64. Dwayne Day, "Cover Stories and Hidden Agendas: Early American Space and National Security Policy," Roger Launius et al. eds., *Reconsidering Sputnik: Forty Years since the Soviet Satellite* (Australia: Harwood Academic, 2000), 167–68.

65. Walter A. McDougall, *The Heavens and the Earth: A Political History of the Space Age* (New York: Basic Books, Inc., 1985), 130.

66. Douglas Aircraft Company, Inc., "Preliminary Design of an Experimental World-Circling Spaceship," Report No. SM-11827, May 2, 1946, *Exploring the Unknown*, 239.

67. A. V. Gross, The Research Institute of Temple University, to Donald A. Quarles, Assistant Secretary of Defense for Research and Development, "Report on the Present Status of the Satellite Problem," August 25, 1953, *Exploring the Unknown*, 268.

68. Dwayne Day, "Cover Stories and Hidden Agendas," *Reconsidering Sputnik*, 170.

69. Kenneth A. Osgood, "Before Sputnik: National Security and the formation of U.S. Outer Space Policy," *Reconsidering Sputnik*, 213.

70. Osgood, 214.

71. Michael Neufeld, "Orbiter, Overflight, and the First Satellite: New Light on the Vanguard Decision," *Reconsidering Sputnik*, 251.

72. Sullivan, *Assault on the Unknown*, 2.

73. "The Feat That Shook the Earth," *Life* (October 21, 1957), 19.

74. Paul Dickson, *Sputnik: The Shock of the* Century (New York: Walker & Co., 2001), 119.

75. McDougall, 200.

76. "Soviet Claiming Lead in Science," *New York Times* (October 5, 1957), 2; "U.S. and Soviet Have Seesawed in Achieving Scientific Advances in Modern Era," *New York Times* (October 6, 1957), 42; and "World Newspapers See Soviet Taking Lead From United States in Space Science," *New York Times* (October 7, 1957), 17.

77. Henry Luce, "Common Sense and Sputnik," *Life* (October 21, 1957), 35.

78. William E. Burrows, *This New Ocean: The Story of the First Space Age* (New York: Random House, 1998), 189.
79. "Unpleasant Information," *Time* (December 9, 1957), 29.
80. "Deterrence & Survival in the Nuclear Age," Security Resources Panel of the Science Advisory Committee (Washington, DC, November 7, 1957).
81. Gretchen J. Van Dyke, "Sputnik: A Political Symbol and Tool in 1960 Campaign Politics," *Reconsidering Sputnik*, 365–400.
82. McDougall, 7.
83. David A. Hollinger, "Science as a Weapon in Kulturkampfe in the United States During and After World War II," *Isis*, 86 (1995), 442; Hollinger, "Free Enterprise and Free Inquiry: The Emergence of Laissez-Faire Communitarianism in the Ideology of Science in the United States," *New Literary History*, 21 (1990), 897–910.
84. Michael Mulkay, *Science and the Sociology of Knowledge* (London: George Allen & Unwin, 1979); Jerry Gaston, "Sociology of Science and Technology," Paul Durbin ed., *The Culture of Science, Technology, and Medicine* (New York: The Free Press, 1980), 474.
85. "Director's Statement," *National Science Foundation Seventh Annual Report for Fiscal Year Ended June 30, 1957* (Washington, DC: National Science Foundation, 1957), x.
86. Statement of Herman Pollack, Acting Director, International Scientific and Technological Affairs of Department of State, Before the House Subcommittee on Science, Research and Development, July 20, 1965, 3. National Archives, RG 307, National Science Foundation, E-39, Office of Antarctic Programs, Central Files, Box 9, 100.51, "House Committee on Science and Astronautics."
87. Mrs. Clare Booth Luce, October 17, 1957 speech at the Alfred E. Smith memorial dinner. Quoted in Lynne L. Daniels, NASA Historical Staff, NASA Historical Note No. 21, "Statements of Prominent Americans on the Opening of the Space Age," July 15, 1963, NASA History Office (NHO), Public Opinion Studies (1957–1975).
88. "Soviet Claiming Lead in Science," *The New York Times* (October 5, 1957), 2; "A Proposal for a 'Giant Leap,'" *Life* (November 16, 1957), 53.
89. Testimony of Livingston T. Merchant, *Review of the Space Program*, House Committee on Science and Astronautics, 86th Cong., 2nd sess., January 20, 22, 25–29, February 1–5, 1960, 3. In awarding its 1957 "Man of the Year" award to Soviet premier Nikita Kruschev, *Time* magazine expressed the awe that could give credence to such propaganda. It recognized that "the U.S. had been challenged and bested in the very area of technological achievement that had made it the world's greatest power." Quoted from "Man of the Year," *Time* (January 6, 1958), 16.
90. Warren Unna, "U.S. Prestige Slip Seen in Poll of 5 Major Allies," *Washington Post* (October 29, 1960).

91. *Increasing the Effectiveness of Western Science*, Foundation Universitaire, Brussels, 1960. National Archives, RG 307, E-1 Office of the Director, 1949–63, Box 50, "President's Science Advisory Committee 1960."

92. Testimony of George V. Allen, *Review of the Space Program*. House Committee on Science and Astronautics, 86th Cong., 2nd sess. Jan 20, 22, 25–29, Feb. 1–5, 1960, 36.

93. "Announcement of the First Satellite," from *Pravda*, October 5, 1957, *Exploring the Unknown*, 330.

94. *Soviet Man in Space* (Moscow, USSR: Foreign Languages Publishing House, 1961), 12.

95. Vannevar Bush, Remarks at Convocation of MIT, "The Essence of Security," December 5, 1949, LOC, Papers of Vannevar Bush, Box 132, file Speech, MIT 12/5/49.

96. Remarks by Frank Stanton at the 1957 International Convention of the Radio Television News Directors Association, November 9, 1957. National Archives, RG 307, E-1, Waterman's Subject Files, Box 33, "Public Information-Requests for Articles and Interviews."

97. Vice President Richard M. Nixon, October 15, 1957. Lynne L. Daniels, "Statements of Prominent Americans on the Opening of the Space Age."

98. NSF Director Alan T. Waterman, Remarks at the Thomas Alva Edison Foundation conference "The Mass Media and the Image of Science," November 6, 1959. NHO, Public Opinion Studies (1957–75).

99. John A. Douglass, "A Certain Future: Sputnik, American Higher Education, and the Survival of a Nation," *Reconsidering Sputnik*, 342–343.

100. Gretchen J. Van Dyke, "Sputnik: A Political Symbol and Tool in 1960 Campaign Politics," *Reconsidering Sputnik*, 396.

101. Statement by L. V. Berkner, June 10, 1960, House Subcommittee on Territorial and Insular Affairs. Library of Congress, Manuscript Collection, Lloyd V. Berkner Papers, Box 17, file Antarctica, 2.

102. H. Guyford Stever, "Our Future in Space," October 20, 1960. Gerald R. Ford Presidential Library, H. Guyford Stever Papers, 1930–90, Box 30, file "MIT-Speeches-Transcripts-Outer Space (2)."

103. "A National Research Program for Space Technology," January 14, 1958. National Archives, RG 307 NSF, E-1 Dr. Waterman's Subject Files, Box 33, file "National Aeronautics and Space Administration."

104. President's Science Advisory Committee, *Introduction to Outer Space* (Washington, DC: US Government Printing Office, March 26, 1958), 1–2.

105. "National Aeronautics and Space Act of 1958," *Exploring the Unknown*, 335.

106. David Binder, "Social Impact," Walter Sullivan ed., *America's Race for the Moon* (New York: Random House, 1962); Dale Carter, *The Rise and Fall of the American Rocket State* (New York: Verso, 1988), 214–222.

107. Office of Program Planning and Evaluation, "The Long Range Plan of the National Aeronautics and Space Administration," December 16, 1959, *Exploring the Unknown*, 404.

108. T. Keith Glennan, Speech before the 4th USAF BMD Symposium on Missiles and Space Technology, Los Angeles, August 24, 1959. National Archives, RG 359 Executive Office of the President, Office of Science and Technology, Subject Files, 1957–62, Box 65, file "Space-Nat'l Aeronautics & Space Admin."

109. Eisenhower "Farewell Address," January 17, 1961.

110. "Report to the President-Elect of the Ad Hoc Committee on Space," January 10, 1961, *Exploring the Unknown*, 421–422.

111. President John F. Kennedy, Special Message to the Congress on Urgent National Needs, May 25, 1961.

112. Livingston T. Merchant, *Review of the Space Program*, 3.

113. "Scientists' Testimony on Space Goals, Senate Committee on Aeronautical and Space Sciences, 88th Cong., 1st sess., June 10 and 11, 1963, 108.

114. T. Keith Glennan speech to the U.S. Air Force, August 24, 1959. National Archives, RG 359, Executive Office of the President Office of Science and Technology, Subject Files 1957–62, Box 65, file "Space-Nat'l Aeronautics & Space Admin."

115. John M. Logsdon, *The Decision to Go to the Moon: Project Apollo and the National Interest* (Cambridge: MIT Press, 1970).

116. *Soviet Man in Space*, 12.

117. James E. Webb, NASA Administrator, and Robert S. McNamara, Secretary of Defense, May 8, 1961, "Recommendations for Our National Space Program: Changes, Policies, Goals," *Exploring the Unknown*, 444.

118. President John F. Kennedy, Special Message to the Congress on Urgent National Needs, May 25, 1961.

119. "The Practical Values of Space Exploration," Staff Study of the House Committee on Science and Astronautics, 87th Cong., 1st sess., August 17, 1961, 3.

120. American Consul in Recife Air Pouch Message to Department of State. National Archives, RG 59 State Department, Central Decimal File 1960–63, Box 1426, file 701.56311/7–2461.

121. Message to the President from Major General Park Chung Hee, Chairman of the Supreme Council for National Reconstruction, July 24, 1961. National Archives, RG 59 State Department, Central Decimal File 1960–63, Box 1426, file 701.56311/7–2461.

122. USIA Research and Reference Service, "Foreign Reaction to Recent U.S. and Soviet Space Activities," February 6, 1964, i. National Archives, RG 59 State Department, SP 10, Box 3146, file "Space Flight & Exploration."

123. Amembassy Beirut airgram to Department of State, "Lebanese Reaction to the Death of Astronauts Grissom, White and Chaffee," February 6,

1967. National Archives, RG 59 State Department, 1967–69 Central Files, Box 2874, file "SP 10–1 US 2/1/67."Keen interest in American astronauts was sometimes a liability in the Middle East. When the crew of Apollo 12 attended a December 1969 event that raised money for reforestation efforts in Israel, several regional newspapers criticized their attendance as "evidence that 'real' U.S. Middle East policy is full support for Israel at the expense of our arab 'friends' in the area." Confidential telegram, Amembassy Jidda to SecState, December 12, 1969. National Archives, RG 59 State Department, 1967–69 Central Files, Box 2872, file "SP 10–1 US 12/1/69."

124. On the tour of Yugoslavia, see: USIS Belgrade to USIA Washington, "Astronaut Tour Stops," November 20, 1969. National Archives, RG 59 State Department, 1967–69 Central Files, Box 2872, file "SP 10–1 US 11–1–69."NASA public affairs officer Brian Duff described astronauts' world tours, including the post-Apollo 11 "Giantstep," Martin Collins et al. eds., *Oral History on Space, Science, and Technology* (Washington, DC: National Air and Space Museum, 1993).

125. Brian Duff memo to Julian Scheer, April 5, 1966. NHO, Public Affairs Box, "Chronological file, 1963–68."

126. Gordon L. Harris, *Selling Uncle Sam* (Hicksville, New York: Exposition Press, 1976), 7.

127. *Educational Programs and Services*, NASA. NHO, Office of Public Affairs Box, file "Educational Programs."

128. On the Spacemobile program, begun in 1961 and then contracted out to several different companies, see: NASA News Release No: 65–188, "NASA to Negotiate with Unitec Corfor Spacemobile Program," June 8, 1965. NASA History Office, Public Affairs box; Vernon Van Dyke, *Pride and Power: The Rationale of the Space Program* (Urbana: University of Illinois Press, 1964), 260.

129. Amembassy, Port-au-Prince airgram to USIA Washington, "Request for All Media Coverage of Haitian on GEMINI Project," November 27, 1965. National Archives, RG 59 State Department, Central Decimal File, 1964–66, Box 3151, file "SP 10 Space & Astronautics, 11/1/65."

130. Amembassy secret air pouch to State Department, "Significance of Outer Space Operations in Relation to the Iranian Political Situation," May 11, 1959. National Archives, RG 59 State Department, 1955–59 Central Decimal File, Box 2772, file 702.02211–2058.

131. L. C. McHugh, "The Why of Space Programs," *America* (May 19, 1962), 267.

132. Paul B. Stares, *The Militarization of Space: U.S. Policy, 1945–1984* (Ithaca: Cornell University Press, 1985).

133. Dwayne A. Day et al. eds., *Eye in the Sky: The Story of the CORONA Reconnaissance Satellite* (Washington, DC: Smithsonian Institution Press, 1998).

134. USUN confidential airgram to Department of State, "Outer Space," November 2, 1965. National Archives, RG 59 State Department, 1964–66, Box 3151, file "SP 10 Space & Astronautics."

135. Robert Seamans, "The Challenge of Space Exploration," *Smithsonian Report for 1961* (Washington, DC: Smithsonian Institution, 1962), 265.

136. Martin J. Collins et al., *Oral History on Space, Science, and Technology* (Washington, DC: Smithsonian National Air and Space Museum, 1993), 261.On Soviet propaganda about US weather and communications satellite, see: Dr. R. W. Porter letter to Lloyd Berkner, April 4, 1960. National Archives, RG 359 Executive Office of the President, Office of Science and Technology, Subject Files 1957–1962, Dummy Box, file "Space-International 1960"; and Amembassy Moscow airgram to State Department, "Soviet Propaganda on Telstar Shifts—now It's an Instrument of the Cold War," August 7, 1962. National Archives, RG 59 State Department, Central decimal File 1960–63, Box 2941, file 901.00/6-162.

137. John Logsdon, *John F. Kennedy and the Race to the Moon* (New York: Palgrave Macmillan, 2010).

138. President Kennedy Address before the 18th General Assembly of the United Nations, September 20, 1963; "Statement by Ambassador Adlai E. Stevenson, U.S. Representative to the United Nations," December 2, 1963. National Archives, RG 59, Central Foreign Policy File 1963, Box 4183, file "SP6 Peaceful Uses of Space UN, 10/1/63."On State Department deliberations over proposing a joint US–USSR lunar landing, see: Confidential memo "In Outer Space It Takes Two To Tango," November 20, 1963; and confidential memo to the Secretary of State, "President Johnson on the UN," November 26, 1963; Robert F. Packard confidential memo to R. Rollefson, "Cooperation with the Soviets in a Joint Expedition to the Moon," November 14, 1963. National Archives, RG 59, Central Foreign Policy File 1963, Box 4183, file "SP-Space & Astronautics, US-USSR."

139. Eileen Galloway, "Organizing the United States Government for Outer Space, 1957–1958," *Reconsidering Sputnik*, 318–321.

140. *Treaty on Principles Governing the Activities of States in the Exploration and Use of Outer Space, Including the Moon and Other Celestial Bodies*, signed January 27, 1967 and entered into force October 10, 1967.

141. Amembassy Fort Lamy airgram to Department of State, "Local Reaction to Apollo 11 Flight," July 31, 1969. National Archives, RG 59 State Department, 1967–69 Central Files, Box 2868, file "SP 10 US 7/3/69."

142. Dulles telegram to Norman Armour, American Embassy, Paris, July 18, 1957. National Archives, RG 59 Department of State, 1955–59 Central Decimal File, Box 2773, file 702.022/4-157.

143. Secret memorandum of conversation, "Defense and State Department Positions Respecting Antarctica," January 30, 1958. National Archives, RG 59 Department of State, 1955–59 Central Decimal File, Box 2773, file 702.022/1–658.

144. "Statement by the President Concerning Antarctica," May 3, 1958.

145. Secret position paper on Antarctic Conference, "Definition of Basic U.S. Purposes in seeking the Treaty, Including Requirements Treaty Must Fulfill in Order to Obtain U.S. Adherence," 5/4/59. National Archives, RG 59 Department of State, 1955–59 Central Decimal File, Box 2773, file 702.022/4–159.

146. Quoted from Article III and Article IX of the Antarctic Treaty, in U.S. Department of State, *Handbook of the Antarctic Treaty System*, 8th edition (April 1994), 12–13.

147. Fracisco Orrego Vicuna, *Antarctic Mineral Exploitation: The Emerging Legal Framework* (Cambridge: Cambridge University Press, 1988), 5–11; and F. M. Auburn, 5–47.

148. "Peace in the Antarctic," *Time* (August 22, 1959); Testimony of Representative John Pillion, *Antarctic Legislation, 1960*, U.S. Congress. House. Subcommittee on Territorial and Insular Affairs, 86th Cong., 2nd sess., June 13–14, 1960, 36.

149. Testimony of the American Legion and National Society of the Daughters of the American Revolution, "Antarctic Legislation, 1960," Hearing Before House Subcommittee on Territorial and Insular Affairs, June 13–14, 1960; President of the National Sojourners to President Eisenhower, November 12, 1959. National Archives, RG 59 Department of State, 1955–59 Central Decimal File, Box 2773, file 702.022/Oct. 1959.

150. Mrs. Charles Snell to Senator Kenneth Keating. National Archives, RG 59 Department of State, 1955–59 Central Decimal File, Box 2773, file 702.022/Oct. 1959.

151. "Antarctica...Now a 'Thaw' in the Coldest Place of All?," *Newsweek* (October 26, 1959), 58.

152. Henry M. Dater, "Organizational Developments in the United States Antarctic Program, 1954–1965," *Antarctic Journal of the United States* (January–February 1966), 26.

153. Joseph Cisco, "The United States Program for Antarctica," *Antarctic Journal of the United States* (January–February 1966), 2–4; Benjamin H. Reed memo to McGeorge Bundy, May 18, 1965. National Archives, RG 59 Department of State, 1964–66 SP1–1 to SP 10, Box 3094, file SCI 11–1 ANT, 1/1/65.

154. Gilbert Dewart, *Antarctic Comrades: An American with the Russians in Antarctica* (Columbus: Ohio State University Press, 1989).

155. Testimony of Livingston Merchant, "Review of the Space Program," House Committee on Science and Astronautics, 86th Cong., 2nd sess., 1960, 18.

156. James E. Heg, "U.S. Interest and Policy in Antarctica," Director's Program Review, Polar Programs, April 9, 1974. National Archives, RG 307 National Science Foundation, E-86 Miscellaneous Reports, 1969–75, Records of the Polar Information Service.

157. Alan Waterman letter to William Finan (Assistant Director of the Bureau of the Budget), May 21, 1959. National Archives, RG 307 NSF, E-39 Office of Antarctic Programs, Central Files, Box 125, file "Bureau of the Budget."

158. United States Antarctic Projects Officer, "Advantage of a Military Nature Derived From Current U.S. Antarctic Operations," 1960. National Archives, RG 313 Records of Naval Operating Forces, Records of the Commander, Admiral's Files, 1955–67, Box 6, file "60–62."

159. "Soviet Interests, Intentions, and Plans for Antarctica," Secret Geographic Intelligence Report, January 27, 1958, Central Intelligence Agency Office of Research and Reports.

160. Confidential "Belgian Operations in Antarctica," November 8, 1960. National Archives, RG 313 Records of Naval Operating Forces, U.S. Naval Support Force, Antarctica, Formerly Classified Records, 1954–72, Box 3, file "1960"; and Classified "Belgian Request for U.S. Assistance in Antarctica," March 5, 1963. National Archives, RG 59 State Department, Central Foreign Policy Files, 1963, Box 4168, file "SCI-Science & Technology ANT-ARG."

161. Harland Cleveland confidential letter to Leland Haworth, October 2, 1964. National Archives, RG 59 State Department, Central Foreign Policy Files, 1964–66, Box 3093, file "SCI 11–1 ANT."

162. Harland Cleveland confidential letter to Leland Haworth, August 6, 1964. National Archives, RG 59 State Department, Central Foreign Policy Files, 1964–66, Box 3093, file "SCI 11–1 ANT"; and Price Lewis, Jr. confidential memo, "Washington Activities," January 13, 1967. National Archives, RG 313 Records of Naval Operating Forces, U.S. Naval Support Force, Antarctica, Formerly Classified Records, 1954–72, Box 2, file "USNSA 66–67."

163. "ANTARCTICA: Inspection and other items," October 23, 1964. National Archives, RG 59 State Department, Central Decimal File, 1964–66 SP 1–1 to SP 10, Box 3093, file "SCI 11–1 Area Research Programs, ANT."

164. T. O. Jones, Head of the NSF Office of Polar Programs, to Alan Waterman, "Antarctic film of the U.S. Navy," January 15, 1963. National Archives, RG 307 National Science Foundation, E-1 Office of the Director, 1949–63, Box 69, Office of Antarctic Programs, 1963; Alan Waterman to Admiral James Reedy, January 29, 1965. National Archives, RG 307 NSF, E-39 Office of Antarctic Programs, Central Files, Box 1, file "Correspondence to the Director."

165. Secret memorandum from Marshall Green to Walter S. Robertson, "U.S. Policy in Antarctica," November 21, 1957. National Archives,

RG 59 Department of State, 1955–59 Central Decimal File, Box 2773, file 702.022/8–657.

166. Confidential State Department "Statement of U.S. Objectives Regarding Antarctica and Courses of Action During the Next Several Years," July 12, 1962. National Archives, RG 313 Naval Operating Forces, formerly Classified Records, 1954–72, Box 1, file "1963"; Secret "Definition of Basic U.S. Purposes in Seeking the Treaty…," May 4, 1959. National Archives, RG 59 State Department, 1955–59 Central Decimal File, Box 2776, file 202.22/4–159.

167. "Current Trends in Soviet Antarctic Interests and Activities," Geographic Support Study, Central Intelligence Agency Office of Research and Reports, July 1964; E. B. Skolnikoff file memo, January 20, 1959. National Archives, RG 359 Executive office of the President, Office of Science and Technology, Subject Files, 1957–62, Dummy Box, file "Antarctica 1959, 60."

168. Classified "Approval for Government Support of a Mountain Climbing Expedition," November 7, 1966. National Archives, RG 59 State Department, 1964–66, Box 3095, file "SCI 11–1 ANT." The Antarctic Policy Group simultaneously refused logistical support to another US expedition that planned to climb the same mountain. This unlucky expedition anticipated climbing the mountain after it had already been "conquered," and it did not have the same credentials and congressional supporters as the American Alpine Club. Three years later, the Antarctic Policy Group refused to provide support to another private expedition—this time a snowmobile expedition headed by the manufacturer of Ski-Doo motor toboggans. See: "Proposed Plaistad Motor Toboggan Expedition to the South Pole," May 21, 1969. National Archives, RG 59 State Department, 1967–69 Central Files Box 2803, file "SCI 11–1 ANT."

169. Alan T. Waterman memo "Proposed Program for Improving Public Understanding of the role of Science," June 20, 1958. National Archives, RG 307 NSF, E-1 Office of the Director, Box 33, file "Public Information;" Clyde Hall memo, "The Role of the Public Information Officer in Government," August 20, 1959, same file; and "Program Objectives of the Public Information Officer," March 1, 1956. National Archives, RG 307 NSF, E-1 Office of the Director, Box 15, file "Public Information."

170. U.S. Congress. House. Subcommittee on Territorial and Insular Affairs. *Antarctic Report, 1962*, 87th Cong., 2nd sess., May 25, 1962, 1.

171. Science News Digest, *Focus on Antarctica* (Hearst Productions, 1960).

172. "Proposal for a Color Motion Picture," 1962. National Archives, RG 307 National Science Foundation, E-39, Office of Antarctic Programs, Central Files, Box 191, 114.6, "Films 1961–1967."

173. Alan Waterman (Head of NSF) letter to McGeorge Bundy (National Security Advisor), June 1, 1962. National Archives, RG 307 NSF, E-39 Office of Antarctic Programs, Central Files, Box 124, file "The White House Office"; and John T. Wilson (Acting Director, NSF) letter to Donald F. Hornig (Director, Office of Science and Technology), June 15, 1967. National Archives, RG 307 NSF, E-39 Office of Antarctic Programs, Central Files, Box 89, file "Midwinter Message File 1961–."

174. President Dwight D. Eisenhower, "Message to Members of All Scientific Expeditions in the Antarctic," June 19, 1959.

175. *U.S. Antarctic Projects Officer*, Summer 1965.

176. President Dwight D. Eisenhower, "Address Before the 15th General Assembly of the United Nations, New York City," September 22, 1960.

177. Message from the President of the United States, *United States Policy and International Cooperation in Antarctica* (Washington, DC: US Government Printing Office, September 2, 1964), 20.

2 The Space and Antarctic Frontiers

1. NASA, *Space: The New Frontier*, EP-6 (Washington, 1966), 2.

2. Vannevar Bush, *Science: The Endless Frontier* (Washington, DC: US Government Printing Office, 1945).

3. "Frontiers of Technology," *Life* (October 7, 1957), 83–84; and "Man of the New Frontier," *Time* (November 16, 1960), 5.

4. Bill announcing the play "Discoveries in the Moon," Philadelphia, 1945. National Air and Space Museum Archives (NASM).

5. Frank H. Winter, "The 'Trip to the Moon' and Other Early Spaceflight Simulations Shows: Part 2," a paper presented at the 47th International Astronautical Congress, October 1996, Beijing, China.

6. Brian Horrigan, "Popular Culture and Visions of the Future in Space, 1901–2001," *New Perspectives on Technology and American Culture* (Philadelphia: American Philosophical Society, 1986), 49–67.

7. Edward E. Hale, "The Brick Moon," *The Atlantic Monthly*, October, November, and December 1869, and February 1870, John M. Logsdon et al. eds., *Exploring the Unknown: Selected Documents in the History of the US Civil Space Program, Vol. I: Organizing for Exploration* (Washington, DC: NASA, 1995), 23–55.

8. Frank H. Winter, "The 'Trip to the Moon' and Other Early Spaceflight Simulations Shows, ca. 1901–1915: Part 1," a paper presented at the 46th International Astronautical Congress, October 1995, Oslo, Norway.

9. Sam Moscowitz, "The Growth of Science Fiction from 1900 to the early 1950s," Frederick I. Ordway III and Randy Lieberman eds., *Blueprint for Space: Science Fiction to Science Fact* (Washington, DC: Smithsonian Press, 1992), 69–82.

10. Michael J. Neufeld, *Von Braun: Dreamer of Space, Engineer of War* (New York: Knopf, 2007).

11. Howard E. McCurdy, *Space and the American Imagination* (Baltimore: Johns Hopkins University Press, 2011), 33–59.

12. Randy Liebermann, "The Collier's and Disney Series," *Blueprint for Space*, 137.

13. Dwayne Day, "Paradigm Lost," *Space Policy* (August 1995), 153–159.

14. David R. Smith, "They're Following Our Script: Walt Disney's Trip to Tomorrowland," *Future* (May 1978).

15. "Man in Space," originally broadcast on March 9, 1955, and then on June 15 and July 29. *Walt Disney Treasures Tomorrowland* (Walt Disney Video, 2004).

16. Donald N. Michael, "The Beginning of the Space Age and American Public Opinion," *Public Opinion Quarterly*, v. 24, n. 4 (Winter 1960), 573–582.

17. Jules Verne, *The Sphinx of the Ice Fields* (1897); James Fennimore Cooper, *Sea Lions* (1849); Edgar Allen Poe, *The Narrative of Arthur Gordon Pym of Nantucket* (1838).

18. "Oasis," *Time* (February 24, 1947), 100. On earlier theories of inhabited Antarctic oases, see William H. Goetzmann, *New Lands, New Men: America and the Second Great Age of Discovery,* (New York: Viking, 1986), 258–264; and Philip I. Mitterling, *America in the Antarctic to 1840* (Urbana: University of Illinois Press, 1959), 67–100.

19. Metro Goldwyn Mayer, *The Secret Land*, 1948. Library of Congress (LOC) Television and Film Archives.

20. Herbert Nichols, "American Interests in the Antarctic," *Christian Science Monitor* (April 17, 1952).

21. Robert E. Byrd speech to Poultry and Egg National Board, Byrd Polar Research Center Archival Program, Papers of Rear Admiral Richard E. Byrd, 054–722.7-2.

22. Beau Riffenburgh, *The Myth of the Explorer: The Press, Sensationalism, and Geographic Discovery* (New York: Belhaven Press, 1993).

23. Among the many hagiographic children's books about Byrd: Robert M. Bartlett, *They Did Something About It* (Freeport: Books for Libraries Press, 1939); Rex Lardner, *Ten Heroes of the Twenties* (New York: Putnam, 1966); Harold and Doris Faber, *American Heroes of the 20th Century* (New York: Random House, 1967); Elliot Robert and Huntley Brown, *Banners of Courage: The Lives of 14 Heroic Men and Women* (New York: Platt & Munk, 1972); and Anne E. Schraff, *American Heroes of Exploration and Flight* (Springfield: Enslow, 1996).

24. Robert E. Byrd, *Alone*, (New York: Ace Books, 1959), Foreword.

25. Thomas R. Henry, *The White Continent: The Story of Antarctica* (New York: William Sloane Assoc, 1950), vii.

26. Rear Admiral Richard E. Byrd letter to Arthur A. Schuck, August 2, 1956. Byrd Polar Research Center Archival Program, Papers of Rear Admiral Richard E. Byrd, 054–540.8-1.

27. "All-Out Assault on Antarctica," *National Geographic* (April 1956).
28. Harvey Sarnat letter to Byrd, April 10, 1954. Byrd Polar Research Center Archival Program, Papers of Rear Admiral Richard E. Byrd, 054–774.22–2.
29. Mark E. Byrnes, *Politics and Space: Image Making by NASA* (Westport: Praeger, 1994), 51.
30. Robert Seamans, "The Challenge of Space Exploration," *Smithsonian Report for 1961* (Washington, 1962), 270.
31. George P. Miller, "Economic Impact of the Space Program," *Signal* (August 1962), 25, 27.
32. Walter A. McDougall, *The Heavens and the Earth: A Political History of the Space Age* (New York: Basic Books, 1985), 202.
33. T. Keith Glennan, "The Challenge of the Space Age," Fort Worth Chamber of Congress, December 8, 1958, NASM Archives, OS-170884.
34. President's Science Advisory Committee (PSAC), "Report of the ad Hoc Panel on Man-in-Space," December 16, 1960, *Exploring the Unknown*, 408.
35. "Report to the President-Elect of the Ad Hoc Committee on Space," January 10, 1961, *Exploring the Unknown*, 422.
36. Harland Manchester, "The Senseless Race to Put Man in Space," *Reader's Digest* (May 1961), 64–67.
37. Alan Waterman letter to Warren Weaver, May 30, 1962. LOC Manuscripts Archive, Papers of Alan T. Waterman, Box 32, file "Warren Weaver."
38. Dr. Philip Abelson, *Scientists Testimony on Space Goals*, Senate Committee on Aeronautical and Space Sciences, 88th Cong., 1st sess., June 10 and 11, 1963, 5; "Nobel Winners Criticize Moon Project," *Washington Post* (May 6, 1963); "Space Program Being Scrutinized by Budget-Minded Congress," *Washington Post* (July 1, 1963).
39. Roger Launius, "Eisenhower, Sputnik, and the Creation of NASA," *Prologue* (Summer 1996): 127–140.
40. President John F. Kennedy, "Special Message to the Congress on Urgent National Needs," May 25, 1961.
41. Sylvia K. Kraemer, "NASA and the Challenge of Organizing for Exploration," Roger D. Launius ed., *Organizing for the Use of Space: Historical Perspectives on a Persistent Issue* (Washington, DC: AAS History Series, 1995), 91; Robert C. Seamans, Jr. and Frederick Ordway, "Apollo Tradition: Object Lesson for the Management of Large-Scale Technological Endeavors," *Interdisciplinary Science Reviews*, v. 2, n. 4 (1977), 276.
42. Brian Balogh, "Reorganizing the Organizational Synthesis: Federal-Professional Relations in Modern America," *Studies in American Political Development* 5 (Spring 1991): 119–172; W. Henry Lambright, *Governing Science and Technology* (New York: Oxford University Press, 1976).

43. Bruce Lewenstein, "NASA and the Public Understanding of Space Science," *Journal of the British Interplanetary Society*, v. 46 (1993), 251–254.

44. Richard E. Horner, Associate Administrator of NASA, April 21, 1960 presentation to the Society of Technical Writers and Editors and the Technical Publishing Society, NASA History Office (NHO), Public Opinion Studies (1957–75).

45. Freeman Dyson, "Human Consequences of the Exploration of Space," Eugene Rabinowitch and Richard S. Lewis eds., *Man on the Moon: The Impact of Science, Technology and International Cooperation* (New York: Basic Books, 1969), 14.

46. Wernher von Braun, "Saturn/Apollo as a Transportation System," *Man on the Moon*, 176.

47. "Antarctica: The Last Frontier," US Antarctic Programs Annual Report, 1956. Byrd Polar Research Center, Alfred N. Fowler Papers, 054–744.21–4, file 7355.

48. John F. Kennedy, "Address of Senator John F. Kennedy Accepting the Democratic Party Nomination for the Presidency of the United States—Memorial Coliseum, Los Angeles," July 15, 1960.

49. Frederick Jackson Turner, "The Significance of History" (1891), in John Mack Faragher ed., *Rereading Frederick Jackson Turner* (New Haven: Yale University Press, 1994), 27; Turner, "The Significance of the Frontier in American History" (1893), *Rereading Frederick Jackson Turner*, 31–32.

50. Turner, "The Problem of the West" (1896), *Rereading Frederick Jackson Turner*, 67.

51. Turner, "The Significance of the Frontier in American History," 33, 38, 59.

52. Frederick Jackson Turner, "Pioneer Ideals and the State University" (1910), *Rereading Frederick Jackson Turner*, 107.

53. Richard Slotkin, *Gunfighter Nation: The Myth of the Frontier in Twentieth-Century America* (New York: Atheneum, 1992), 22.

54. Turner, "The Significance of the Frontier in American History," 47.

55. Slotkin, *Gunfighter Nation*, 22.

56. John Mack Faragher, "Introduction," *Rereading Frederick Jackson Turner*, 8.

57. Turner, "Pioneer Ideals and the State University," *Rereading Frederick Jackson Turner*, 114, 117.

58. Bush, *Science—The Endless Frontier*, Letter of Transmittal.

59. Remarks by Dr. Glenn T. Seaborg, "A Scientific Society—The Beginnings," at the American Association for the Advancement of Science, December 27, 1961. National Archives, RG 326, E-24, Box 8, file "Speeches, 1961, part 1."

60. Alan T. Waterman, "The Crisis in Science Education: The Role of the Federal Government," December 29, 1955, LOC, Alan T. Waterman Papers, Box 47, "The Role of the Federal Government" file.

61. Paul Boyer, *By the Bomb's Early Light: American Thought and Culture at the Dawn of the Atomic Age* (Chapel Hill: University of North Carolina Press, 1994); Spencer Weart, *Nuclear Fear: A History of Images* (Cambridge: Harvard University Press, 1988).

62. Seaborg, "A Scientific Society—The Beginnings."

63. Press Release "Atoms for Peace" Speech, December 8, 1953. Dwight Eisenhower Presidential Library, DDE's Papers as President, Speech Series, Box 5, United Nations Speech 12/8/53.

64. Robert Collins, *More: The Politics of Economic Growth in Postwar America* (New York: Oxford University Press, 2000).

65. Lloyd V. Berkner, *The Scientific Age: The Impact of Science on Society* (New Haven: Yale University Press, 1964), 4–5.

66. Secretary Christian A. Herter, "United States Foreign Policy Under the Eisenhower Administration, 1953–1961," *Department of State Bulletin* (January 30, 1961), 144.

67. Statement by Lloyd V. Berkner, House Subcommittee on Territorial and Insular Affairs, June 10, 1960. LOC, Lloyd V. Berkner Papers, Box 21, Statement on Antarctica 10 June 1960 file.

68. Gerard Piel, *Science in the Cause of Man* (New York: Alfred A. Knopf, 1961), 188.

69. Vice President Richard M. Nixon, October 15, 1957. Lynne L. Daniels, NASA Historical Staff, "Statements of Prominent Americans on the Opening of the Space Age," July 15, 1963, NASA Historical Note No. 21. NHO, Public Opinion Studies (1957–75).

70. Richard M. Nixon, presidential candidate press release, "The Scientific Revolution," September 8, 1960, National Archives, RG 307 National Science Foundation, Entry-39 Office of Antarctic Programs Central Files, Box 124, 110.1, "The White House Press Release."

71. Edward H. White, II, "Democracy and Space," *Space World* (September 1964), 32.

72. T. Keith Glennan speech to the US Air Force, August 24, 1959. National Archives, RG 359, Executive Office of the President Office of Science and Technology, Subject Files 1957–62, Box 65, file "Space-Nat'l Aeronautics & Space Admin."

73. Rear Admiral George Dufek, "What We've Accomplished in Antarctica," *National Geographic* (September 1957), 342.

74. United States Information Agency, *Power for Continent Seven*, 1963, National Archives, Motion Picture, Sound, and Video Branch, NC3-306-77-7.

75. Joseph M. Dukert, *This Is Antarctica* (New York: Coward, McCann & Geoghegan, 1965), 173.

76. Albert Crary, "Where Is Science Taking Us?," *The Saturday Review* (March 3, 1962), 40.

77. Dr. Wernher von Braun, "A Space Man's Look at Antarctica," *Popular Science* (May 1967); Samuel Matthews, "Antarctica: Icy Testing Ground for Space," *National Geographic* (October 1968), 570.

78. "NASA Officials Visit Antarctica," *Antarctic Journal of the United States* (March–April 1967), 52.

79. "Key Guide to Outer Space," *This Week Magazine* (October 1958), 19.

80. James Clavell, "The First Woman on the Moon," *Men Into Space*, United Artists Corporation, broadcast October 14, 1959. State Historical Society of Wisconsin, Archives Division, US Mss 99AN.

81. CBS, *The 21st Century*: "To the Moon," February 5, 1967; and "Mars and Beyond," March 19, 1967. LOC Television and Film Archives.

82. Walter Froelich, "A Look Into the Future of Space Exploration," *USIS Feature* December 1967. Lyndon B. Johnson Presidential Library, Office Files of Frederick Panzer, Box 181; and "America's Next Decades in Space: The Post-Apollo Program, Directions for the Future," NASA Report to the President's Space Task Force, September 1969.

83. Dr. Edward C. Welsh, Executive Secretary, National Aeronautics and Space Council, "Returns from the Space Dollar," 11 October 1966. NASM Archives, OS-170884–44.

84. Homer Newell, "The Impact of Space Activities on the Community of Nations," NASA Release, April 14, 1962. National Archives, RG 255 National Aeronautics and Space Administration, E-86 Records of Dr. Homer Newell, Associate NASA Administrator, Box 94.

85. "Why Spend $20 Billion to Go to the Moon?," *US News & World Report* (July 3, 1961), 60.

86. "This Is NASA," Washington, DC: NASA, EP-22, December 1968. NASM Archives, ON-010110–01/02; James E. Webb, "Accelerated Space Exploration: An Imperative for Americans," February 18, 1962, NASA News Release No. 62–34.

87. "United States Policy and International Cooperation in Antarctica, Message from the President of the United States," Committee on Foreign Affairs, September 2, 1964, 20.

88. Walter Sullivan, "Why Antarctica is Being Explored," *New York Times Magazine* (March 6, 1960), 20.

89. National Science Foundation, *Seventh Annual Report for the Fiscal Year Ended June 30, 1957* (Washington, DC: Government Printing Office, 1957), 20.

90. Roger Caras, *Antarctica: Land of Frozen Time* (New York: Chilton Books, 1962), 153.

91. National Science Foundation, *Introduction to Antarctica* (Washington, DC: NSF, 1969), 46.

92. W. W. Rostow, *The Stages of Economic Growth: A Non-Communist Manifesto* (Cambridge: Cambridge University Press, 1960).

93. Michael E. Latham, *Modernization as Ideology: American Social Science and "Nation Building" in the Kennedy Era* (Chapel Hill: University of North Carolina Press, 2000), 6; Nils Gilman, *Mandarins of the Future: Modernization Theory in Cold War America* (Baltimore: Johns Hopkins University Press, 2003).

94. Bryce Nelson, "Science in Antarctica: Problems and Opportunities," *Science* (January 26, 1968), 407.

95. Harold M. Schmeck, Jr., "From Dog Sled to Ski-Plane: In Fifty Years Much Has Changed at the South Pole—Except the Cold," *New York Times Magazine* (February 25, 1962), 24.

96. National Science Foundation, *Ninth Annual Report for the Fiscal Year Ended June 30, 1959* (Washington, DC: Government Printing Office, 1959), xi.

97. Ladd Plumley, "Facing Up to Space," lecture before Harvard Business School Club of New York, October 29, 1962. Hagley Museum and Library, 1960 Series 1, Box 21.

98. David Binder, "Social Impact," Walter Sullivan ed., *America's Race for the Moon* (New York: Random House, 1962), 121–133; Charlie Evans, "NASA Boom Still Glows," *Houston Chronicle* (March 7, 1965).

99. "Florida Rides a Space-Age Boom," *National Geographic* (December 1963); Dave W. Lang, "Space Dollars," NASA Fact Sheet #259, May 31, 1964. Hagley Library.

100. John W. Finney, "NASA May Leave Its Alabama Base," *The New York Times* (October 24, 1964); "NASA Chief Critical of Huntsville's National Image," *The Huntsville Times* (January 19, 1967); "The 'Image' vs. the 'Reality,'" *Huntsville Times* (January 20, 1967).

101. W. Henry Lambright, *Governing Science and Technology* (New York: Oxford University Press, 1976), 156.

102. George P. Miller, "Economic Impact of the Space Program," *Signal* (August 1962), 25, 27.

103. Leonard Sayles, "James Webb at NASA," *Society* (September/October 1992), 64.

104. McDougall, *The Heavens and the Earth*, 6; Staff Study, "Future National Space Objectives," House Subcommittee on NASA Oversight, 89th Cong., 2nd sess., 1966, 3.

105. Robert C. Seamans, Jr. and Frederick Ordway, "Apollo Tradition: An Object Lesson for the Management of Large-Scale Technological Endeavors," *Interdisciplinary Science Reviews*, v. 2, n. 4 (1977), 270–304.

106. James Webb, "NASA As An Adaptive Organization," John Diebold Lecture, Harvard University, September 30, 1968. Lyndon B. Johnson Presidential Library, Administrative History of the National Aeronautics and Space Administration, Box 3; James E. Webb, *Space-Age Management: The Large-Scale Approach* (New York: McGraw Hill, 1969); W. Henry Lambright, *Powering Apollo: James E. Webb of NASA* (Baltimore: Johns Hopkins University Press, 1995).

107. Glenn T. Seaborg and William R. Corliss, *Man and Atom: Building a New World through Nuclear Technology* (New York: E.P. Dutton, 1971), 174.

108. "Construction: The Earth Mover," *Time* (May 3, 1954), 86.

109. Seaborg, *Man and Atom*, 174.

110. James T. Patterson, *Grand Expectations: The United States, 1945–1974* (New York: Oxford University Press, 1996), 317.

111. CBS, *The 21st Century*: "To the Moon," February 5, 1967; "Mars and Beyond," March 19, 1967; "Conquering the Sea," April 30, 1967; "The Mighty Atom," May 14, 1967; "Can We Live to Be 100," March 24, 1968; and "Can We Control the Weather," September 22, 1968. LOC Television and Film Archives.

112. "Man of the Year, Twenty Five and Under," *Time* (January 6, 1967), 18.

113. M. Philip Copp, *The Atomic Revolution* (Washington, DC: Atomic Energy Commission, 1957). Hagley Library, Accession 1410, Box 2.

114. Richard G. Hewlett and Jack M. Holl, *Atoms for Peace and War, 1953–1961: Eisenhower and the Atomic Energy Commission* (Berkeley: University of California Press, 1989), 528–530; Michael L. Smith, "'Planetary Engineering': The Strange Career of Progress in Nuclear America," John L. Wright ed., *Possible Dreams: Enthusiasm for Technology in America* (Dearborn: Henry Ford Museum and Greenfield Village, 1992), 110–123.

115. Michael L Smith, "Making Time: Representations of Technology at the 1964 World's Fair," T. J. Jackson Lears and Richard Wrightman Fox eds., *The Power of Culture: Critical Essays in American History* (Chicago: University of Chicago Press, 1993), 223–244.

116. *Let's Go to the Fair and Futurama*, New York World's Fair, 1964–65. Hagley Library, Pam 89.631.

117. *Official Guide, New York World's Fair, 1964/1965* (New York: Time Inc., 1964), 77, 90, 204.

118. Roland Marchand, "Corporate Imagery and Popular Education: World's Fairs and Expositions in the United States, 1893–1940," David E. Nye and Carl Pederson eds., *Consumption and American Culture* (Amsterdam: VU University Press, 1991), 18–33.

119. US Department of Commerce, "United States Science Exhibit, Seattle World's Fair, Final Report," 1962, 3; "Century 21 Exposition." National Archives, RG 306, Records Relating to the 1967 Universal and International Exhibition in Montreal-Background Files, 1964–76, Box 161, file "Seattle World's Fair."

120. Norman K. Winston, "Challenge to Greatness: The American Journey, The Exhibition Content for the United States Pavilion, New York World's Fair 1964–1965," Office of the Commissioner, August 26, 1963. National Archives, RG 306, Records Relating to the 1967 Universal and International Exhibition in Montreal- Background Files, 1964–76, Box 161, file "New York World's Fair."

121. *United States Policy and International Cooperation in Antarctica: Message from the President of the United States* (Washington, DC: US Government Printing House, September 2, 1964), 20.

122. John McCone, Chairman US Atomic Energy Commission, "The Influence of Nuclear Technology on Rockets and Space," November 18, 1959, 2. National Archives, RG 326 Record's of the Atomic Energy Commission, Records of the Office of Public Information, e-24 Copies of Speeches of AEC Officials, 1947–74, Box 7, file AEC Speeches, 1959, part 1.

123. Dr. Richard Engler, *Atomic Power in Space: A History* (Washington, DC: Department of Energy, March 1987), 8–9.

124. Confidential telegram from the Department of State to the US Mission, Geneva, May 29, 1964. National Archives, RG 59 Department of State, Central Decimal File, 1964–66, SP1–1 to SP 10, Box 3147, file "SP 10 Space Flight & Exploration."

125. Gordon L. Harris, *Selling Uncle Sam* (Hicksville, New York: Exposition Press, 1976), 17.

126. Engler, *Atomic Power in Space*, 26.

127. President's Science and Advisory Committee, "Report of the Ad Hoc Panel on Man-in-Space," December 16, 1960, *Exploring the Unknown*, 1995, 409.

128. "Trip to Mars is Linked to Atom Rocket," *New York Herald Tribune* (July 19, 1959).

129. Dr. Arthur Kantrowitz, "The Importance of Motivation in Pacing Future Space Exploration," Remarks at the dedication of the Avco Research Center, May 14, 1959, NASM Archives, file OS-170884-30.

130. Hugh Dryden, "Looking Ahead to Space," The 1964 Bicentennial Martial Wood Lecture, Brown University, October 21, 1964. Published in *Speaking of Space and Aeronautics*, vol. 1, no. 4 (Washington, DC: NASA, March 1965); NASA's "Summary Report: Future Programs Task Group, January 1965," *Exploring the Unknown*, 1995, 473–489.

131. Edward Teller, "The Future Is Yours," Remarks at the Nuclear Space Seminar, Amarillo, Texas, August 27–28, 1962. LOC, Lloyd V. Berkner Papers, Box 37, file "Space Technology and You."

132. Seaborg, *Man and Atom*, 220.

133. Walter Sullivan, *Quest for a Continent* (New York: McGraw-Hill, 1957), 352.

134. "Preliminary Study of Nuclear Power Plants for Operation DEEPFREEZE," September 12, 1955. National Archives, RG 313 Records of Naval Operating Forces, Box 6, file "1955"; Admiral Carney to J.A. Hutchinson, November 2, 1956. Operational Archives Branch, Naval Historical Center, Naval Support Force, Arctic & Antarctic, Admiral George Dufek, 1956–59.

135. "Nuclear Power Comes to Antarctica," *Bulletin of the US Antarctic Projects Officer* (March 1962), 10.

136. Owen Wilkes and Robert Mann, "The Story of Nukey Poo," *Bulletin of the Atomic Scientists* (October 1978), 34.

137. Peter Briggs, *Laboratory at the Bottom of the World* (New York: David McKay, 1970), 148–149.

138. Proposed letter from Alan Waterman to Glenn Seaborg. National Archives, RG 307 NSF, E-39 Office of Antarctic Programs, Central Files, Box 88, file "Nuclear Reactors for Antarctica 63"; and T. O. Jones to Alan Waterman, "Nuclear Reactors in the Antarctic…," June 7, 1960. National Archives, RG 307 NSF, E-1 Waterman's Subject Files, Box 42, file "Office of Special International Programs."

139. Melvin A. Rosen et al. (AEC Division of Reactor Development) "Compact Reactor Development for Ground Nuclear Power Applications," June 1961. National Archives, RG 326, E-24, Box 8, file "Speeches, 1961, part 2"; US Naval Support Force, Antarctica, *Support for Science: Antarctica*, History and Research Division, 1965, 37.

140. W. G. Shafer, US Navy, "Five Years of Nuclear Power at McMurdo Station," *Antarctic Journal of the United States* (March–April 1967), 39; Alan Waterman to David Bell, Director of the Bureau of the Budget, February 19, 1962. National Archives, RG 307 NSF, E-1 Waterman's Subject Files, Box 57, file "Office of Antarctic Programs 1962"; Deputy Secretary of Defense Roswell Gilpatric to Alan Waterman, May 1, 1963. National Archives, RG 307 NSF, E-39 Office of Antarctic Programs, Central Files, Box 88, file "Nuclear Reactors for Antarctica 63."

141. Draft letter from Commander, Naval Support Antarctica to Commander, Naval Facilities Engineering Command, "The PM-3A Waste Water Program," August 15, 1968. National Archives, RG 307 NSF, E-39 Office of Antarctic Programs, Central Files, Box 88, file "Nuclear Power for Antarctica"; United States Information Agency, *Power for Continent Seven*, 1963, National Archives, Motion Picture, Sound, and Video Branch, NC3–306–77–7.

142. "Fact Sheet on PM-3A to be Released to Press in Response to Queries," September 25, 1972. National Archives, RG 307 NSF, E-120 Central Subject Files, 1976–87, Division of Polar Programs, Box 3, file "McMurdo Station (Nuclear Power Plant)."

143. Chunglin Kwa, "The Rise and Fall of Weather Modification: Changes in American Attitudes toward Technology, Nature, and Society," Clark Miller and Paul Edwards eds., *Changing the Atmosphere: Expert Knowledge and Environmental Governance* (Cambridge: MIT Press, 2001), 135–165; W. N. Hess ed., *Weather and Climate Modification* (New York: John Wiley, 1974); Robert Fleagle ed., *Weather Modification: Science and Public Policy* (Seattle: University of Washington Press, 1969).Congressman James Fulton (R, PA) feared a Cold War weather-control gap: *Weather Modification*, Hearing before the House Committee on Science and Astronautics, 86th Cong., 1st sess., February 16, 1959, 30.

144. W. Henry Lambright, "Government and Technological Innovation: Weather Modification as a Case in Point," *Public Administration Review* (January/February 1972), 3–4.

145. Homer E. Newell, "A Recommended National Program in Weather Modification, A Report to the Interdepartmental Committee for Atmospheric Sciences" (Washington, DC, October 1, 1966), 11–28.In its year-end report, the NSF announced that weather modification "had now reached a point where rational programs could be initiated." National Science Foundation, *Sixteenth Annual Report for the Fiscal Year Ended June 15, 1966*, xv.

146. CBS television, "Can We Control the Weather?," *The 21st Century* (September 22, 1958). LOC Television and Film Archives.

147. *Bulletin of the US Antarctic Projects Officer* (October 1963), 33.

148. *Antarctica: Coldest Continent* (Marble-Gibson Production, 1968).

149. Henri Bader memo, "Creation of a New Oasis in Antarctica," January 10, 1967. National Archives, RG 307 NSF, E-39 Office of Antarctic Programs, Central Files, Box 7, file 100.23.

150. Joseph M. Dukert, *This is Antarctica*, 153.

151. *Let's Go to the Fair and Futurama*, New York World's Fair, 1964–65.

152. Hermann Oberth, *Man Into Space: New Projects for Rocket and Space Travel* (New York: Harper & Brothers, 1957), 107; Ernest Haussman, "The World of 1971," *Space World* (March 1961), 30.

153. The Honorable Lionel Van Deerlin in General Dynamics, Astronautics Division, *2063 A.D.* (July 1963), 40.

154. The NHO has a file "Project 'Brilliant', 'Moonlight', 'Reflector'" the declassified documents: Edward Grey memo to NASA Deputy Associate Administrator, "Project Moonlight," March 18, 1966; Maurice Raffensperger letter to Francis Williams, April 15, 1966; NASA Deputy Administrator letter to DOD Director of Defense Research and Engineering, January 5, 1967; and "Administrator's Back-up Book, New Item—Project ABLE," July 24, 1967.

155. NASA, "Some Thoughts on the Next Decade of Space Exploration," November 21, 1958. National Archives, RG 307 NSF Office of the Director, 1949–63, Box 33, file "N.A.S.A.," 4; *NASA...Spearhead to Space* (Washington, DC: NASA, 1960).

156. Space Science Board Memorandum SSB-247," December 1, 1961. National Archives, RG 307 NSF, E-1 Waterman's Subject Files, Box 53, file "Space Science Board, January-June 1961," 7; Morris Neiburger, "Utilization of Space Vehicles for Weather Prediction and Control," Simon Ramo ed., *Peacetime Uses of Outer Space* (New York: McGraw-Hill, 1961), 155.

157. *Statements by Presidents of the United States on International Cooperation in Space, A Chronology: October 1957–August 1971*, Senate Committee on Aeronautical and Space Sciences, September 24, 1971, 31.

158. "Remarks to Apollo 11 Astronauts Aboard the U.S.S. Hornet Following Completion of Their Lunar Mission, July 24, 1969," *Public Papers of the Presidents, Richard Nixon 1969* (Washington, DC: US Government Printing Office, 1970), 645.
159. "On Courage in the Lunar Age," *Time* (July 25, 1969).
160. David M. Wrobel, *The End of American Exceptionalism: Frontier Anxiety From the Old West to the New Deal* (University Press of Kansas, 1993), 37.
161. Theodore Roosevelt, *The Winning of the West* (Greenwich: Fawcett Pub., 1963), xi.
162. Roosevelt, *The Winning of the West*, 31.
163. Roosevelt, *The Winning of the West*, 21.
164. Theodore Roosevelt, "On Motherhood," speech given to National Congress on Women, March 13, 1905.
165. Slotkin, *Gunfighter Nation*, 34.
166. Gail Bederman, *Manliness and Civilization: A Cultural History of Gender and Race in the United States, 1880–1971* (Chicago: University of Chicago Press, 1995), 186.
167. Theodore Roosevelt, "The Strenuous Life," Speech before the Hamilton Club, Chicago, April 10, 1899.
168. David M. Potter, *People of Plenty: Economic Abundance and the American Character* (Chicago: University of Chicago Press, 1954); John Kenneth Galbraith, *The Affluent Society* (Boston: Houghton Mifflin, 1958).
169. Warren Susman, "Did Success Spoil the United States? Dual Representations in Postwar America," Lary May ed., *Recasting America: Culture and Politics in the Age of Cold War* (Chicago, 1989), 19–37.
170. David Reisman et al., *The Lonely Crowd: A Study of the Changing American Character* (New Haven: Yale University Press, 1950), 41.
171. William H. Whyte, *The Organization Man* (New York: Simon & Schuster, 1956); Vance Packard, *The Hidden Persuaders* (New York: McKay, 1957); Daniel Boorstin, *The Image: Or What Happened to the American Dream* (New York: Atheneum, 1961).
172. Reinhold Niebhur, "After Sputnik and Explorer," *Christianity and Crisis*, v. XVIII, no. 4 (March 17, 1958), 30.
173. "Excerpts from the House Report on Space Policy, *New York Times* (January 11, 1959), F1.
174. George Miller, "Economic Impact of the Space Program," 27.
175. Harry Stine, quoted in Dr. C. C. Furnas, "Why Did US Lose the Race? Critics Speak Up," *Life* (October 21, 1957), 23.
176. Lynne L. Daniels, NASA Historical Staff, "Statements of Prominent Americans on the Opening of the Space Age," July 15, 1963, NASA Historical Note No. 21. NHO, Public Opinion Studies (1957–75).
177. Erwin D. Canham, President, Chamber of Commerce of the United States, "Business Responsibility in the World Today, Keynote

Address, 48th Annual Meeting, May 1960; Canham, "The World and National Outlook," October 6, 1959. Hagley Library and Museum, 1960, Series 1, Box 21.

178. Billy Graham letter to President Dwight D. Eisenhower, December 2, 1957. NHO, file "Impact, Public Opinion: Sputnik," 006737.

179. T. J. Jackson Lears, "Mass Culture and Its Critics," Mary Cupiec Clayton et al. eds., *Encyclopedia of American Social History* (New York: Charles Scribner's Sons, 1993), 1600.

180. *Goals for Americans: The Report of the President's Commission on National Goals* (New York: Prentice-Hall, 1960), 23; *Prospect for America: The Rockefeller Panel Reports* (Garden City: Doubleday & Co., 1961).

181. Henry Luce, "Common Sense and Sputnik," 35.

182. "Director's Statement," *National Science Foundation Eighth Annual Report for the Fiscal Year Ended June 30, 1958* (Washington, DC: US Government Printing House, 1958), xi, xiv; NSF Director Alan Waterman, "The Mass Media and the Image of Science," Remarks at the Thomas Alva Edison Foundation Conference, November 6, 1959. NHO, Public Opinion Studies (1957–75).

183. J. R. Killian, "Making Science a Vital Force in Foreign Policy," September 23, 1960. National Archives, RG 359 Executive Office of the President, Office of Science and Technology, Subject Files 1957–1962, Box 65, file "Ad Hoc Man-in-Space Panel."

184. President's Science Advisory Committee, *Introduction to Outer Space* (Washington, DC: US Government Printing Office, March 26, 1958), 1.

185. Dr. Dorothy M. Simon, "Mankind in the Space Age," May 14, 1959. NASM, OS-170884–30; Seamans, "The Challenge of Space Exploration," 270.

186. "Betting on a Good Thing," *Hartford Times* (May 16, 1963).

187. A. R. Hibbs, "Must We Rationalize Astronautics?," *Space Digest* (July 1961), 78.

188. President John F. Kennedy, "Midwinter Message," *Bulletin, US Antarctic Projects Officer* (June 1963), 1; President Lyndon B. Johnson, "Midwinter Message," *Bulletin, US Antarctic Projects Officer* (Summer 1965), 1.

189. Rear Admiral David Tyree, "Operation Deep Freeze 61," *Bulletin, US Antarctic Projects Officer* (June 1961), 2; "In Memorium," *Bulletin, US Antarctic Projects Officer* (November 1961), 1; "In Memorium," *Bulletin, US Antarctic Projects Officer* (March–April 1966), 39.

190. President's Science Advisory Committee, *Introduction to Outer Space*, 11.

191. "The Spirit of Adventure," *Life* (November 24, 1958), 34.

192. Marcel LaFollette, *Making Science Our Own: Public Images of Science, 1910–1955* (Chicago: University of Chicago Press, 1990), 106.

193. Merle Tuve, "Is Science Too Big for the Scientist?," *Saturday Review* (June 6, 1959), 49.

194. NBC, "Assault on Antarctica," January 18, 1960. LOC Television and Film Archives.

195. "To the Moon and Back," *Life* special edition (1969).

196. "You and Space," The Boeing Company, Aerospace Group (1972), 3–4. NASM, OS-170950-01.

197. John Willard Ward, "The Meaning of Lindbergh's Flight," *American Quarterly* v. 10, n. 1 (Spring 1958), 3–16.

198. "On Courage in the Lunar Age," *Time* (July 25, 1969), 19.

199. Michael Smith, "Selling the Moon: The US Manned Space Program and the Triumph of Commodity Scientism" in *The Culture of Consumption: Critical Essays in American History, 1880–1980* (New York, 1983), 200.

200. James Earl Rudder, "Preparation of the Whole Man," Nuclear Space Seminar, Amarillo, Texas, August 1962. LOC, Lloyd V. Berkner Papers, Box 37, file "Space Technology and You."

201. "Remarks on Presenting the Presidential Medal of Freedom to Apollo 13 Mission Operations Team in Houston," April 18, 1970. *Public Papers of the Presidents, Richard Nixon, 1970* (Washington, DC: Office of Federal Register, 1971), 367.

202. Material on NASA public affairs operations is available in: National Archives, RG 59, 1967–69 Central Files, Box 2872; NHO, Public Affairs Box; Brian Duff, "Storytellers in Space: NASA and the Media," *Space: Discovery and Exploration* (Washington, 1993); Gordon L. Harris, *Selling Uncle Sam* (Hicksville, 1976).

203. Representative Lionel Van Deerlin March 30, 1966 letter to Dave Bunn. Lyndon B. Johnson Library, Papers of Lyndon Baines Johnson President, 1963–69, Subject File, Box 295, File FG 260 8/12/65.

204. Julian Scheer memo to Robert Seamans, July 9, 1964. NHO, file "Julian Scheer 001903"; National Capital Area Council, Boy Scouts of America, *Explorer Space Program*" 1962, NHO, Office of Public Affairs box, file "Educational Programs"; Boy Scouts of America, *Space Exploration*, New Brunswick, New Jersey, 1966.

205. Hiden T. Cox (NASA Assistant Administrator for Public Affairs) memo "Policy on Astronaut Appearances and Showing of MA-6 Capsule," NHO, Office of Public Affairs box, file "Correspondence."

206. John Glenn, Jr., *"P.S. I listened to your heartbeat": Letter to John Glenn* (Houston, 1964), 92, 3.

207. Alan Waterman, "Our Swiftest Diligence," Loyola University Commencement Speech, June 7, 1962. LOC, Alan T. Waterman Papers, Box 46, "Our Swiftest Diligence" file.

208. Lillian Levy, "Air Force Nurses Preparing For Their Role in Space Age," *Washington Evening Star* (November 12, 1959), 16.

209. Julian Scheer letter to Miss Janet Warnke, September 16, 1963. NHO, file "Julian Scheer 001903."

210. "Woman Space Pioneer," *Philadelphia Inquirer* (January 15, 1959); "Womanned Flight," *Christian Science Monitor* (February 13, 1959).

211. "Qualifications for Astronauts," House Committee on Science and Astronautics, 87th Cong., 2nd sess., July 17 and 18, 1962; Margaret A. Weitekamp, *The Right Stuff, the Wrong Sex: The First Women in Space Program* (Baltimore: Johns Hopkins University Press, 2005); Joan McCullough, "13 Who Were Left Behind," *Ms.* (September 1973), 41–45.

212. "Women are Different," *Time* (June 28, 1963).

213. Julian Scheer letter to Miss Virginia Allan, President of the National Federation of Business and Professional Women's Clubs, Inc., August 28, 1963. NHO, file "Julian Scheer, 001903."

214. Quoted in Ena Naunton, "A Woman in Outer Space? Good Idea-But Still a Joke," *The Miami Herald* (May 21, 1972), 17.

215. James Clavell (scriptwriter), "First Woman on the Moon," *Men into Space*, United Artists, October 14, 1959. State Historical Society of Wisconsin, Archives Division, United Artist Corporation, US Mss 99AN.

216. 1964 draft letter, T. O. Jones to Rear Admiral James R. Reedy, USN. National Archives, RG 307 National Science Foundation, E-39 Office of Antarctic Programs, Central Files, Box 59, file "Women in Antarctica."

217. Philip Smith memo to T. O. Jones, "Meeting with Captain M.W. Nicholson," May 21, 1964. National Archives, RG 307 National Science Foundation, E-39 Office of Antarctic Programs, Central Files, Box 59, file "Women in Antarctica."

218. 1997 interview with retired Rear Admiral Lloyd Abbot, commander of the US Navy's Antarctic operations, 1967–69; 1997 interview with Jack Reniere, NSF Public Affairs Officer.

219. Barbara Land, *The New Explorers: Women in Antarctica* (New York: Dodd, Mead & Co., 1981), 214; Elizabeth Chipman, *Women on the Ice: A History of Women in the Far South* (Melbourne: Melbourne University Press, 1986).

220. Alan Waterman, Loyola University Commencement, "Our Swiftest Diligence," June 7, 1982. LOC, Alan T. Waterman papers, Box 46, file "Our Swiftest Diligence," 8.

221. "Science in Antarctica," *Science* (January 26, 1968), 409.

222. L. C. McHugh, "The Why of the Space Programs," *America* (May 19, 1962), 268.

223. This common metaphor was the title of an enormously popular 1955 photography exhibit, whose coffee-table catalogue ultimately sold over four million copies. See *The Family of Man* (New York: Maco Magazine Corp., 1955); Eric J. Sandeen, "The Family of

Man at the Museum of Modern Art: The Power of the Image in 1950s America," *Prospects* (1986): 367–391.

224. "Destination 'Moon,'" *Ebony*, v. 13 (July 1958), 120–124; "Space Doctor for the Astronauts," *Ebony*, v. 17 (April 1962): 35–39; "Angel of Mercy to the Astronauts," *Ebony*, v. 21 (June 1966), 49–52.

225. "Statement by George Mueller," July 24, 1969. Gerald R. Ford Library, H. Guyford Stever Papers, 1930–90, Box 66, file "NASA (1967–70)."

226. American Consul in Recife Air Pouch Message to Department of State. National Archives, RG 59 State Department, Central decimal File 196–63, Box 1426, file 701.56311/7–2461.

3 Antarctica and the Greening of America

1. President Gerald R. Ford, "Reaching for the Unknown," July 1, 1976; *The American Adventure: The Bicentennial Messages of Gerald R. Ford*, July 1976, 1, 3.

2. Press Release, Office of the White House Press Secretary, "Telephone Conversation between the President and Dr. James C. Fletcher, Administrator, National Aeronautics and Space Administration at the Kennedy Space Center," June 14, 1976. Gerald R. Ford Presidential Library, Glenn R. Schleede Files, Box 21, file "NASA, 1976: General, January-June."

3. "World of Century 21, Official Guide Book, Seattle World's Fair 1962," Hagley Library, Pam 89.631; "Let's Go to the Fair and Futurama, New York World's Fair, 1964–1965," Hagley Library, Pam 89.631.

4. John R. Stiles, White House, "Preliminary Concept of a U.S. Bicentennial Exposition of Science and Technology to Be Held at Cape Canaveral, Florida, June–September 1976," July 10, 1975. Ford Library, Kenneth A. Lazarus Files, 1974–77, Box 14, File "FG 370 American Revolution Bicentennial Administration."

5. Yaron Ezrahi, "The Political Resources of American Science," *Science Studies* (1971), 124.

6. Robert Heilbroner, *An Inquiry into the Human Prospect* (New York: W.W. Norton, 1974), 16.

7. "Science, Progress and Freedom," Address by Vice President Nelson A. Rockefeller before the American Association for the Advancement of Science, Boston, Massachusetts, February 23, 1976. Ford Library, Glenn R. Schleede Files, Box 41, File "Sci and Tech Policy, 1976: Speech Material."

8. William S. Banowsky, "Whatever Happened to the American Dream?, Freedom Is Indivisible," speech before the Philadelphia Society, Los Angeles, California, January 31, 1976. Ford Library, Glenn R. Schleede Files, Box 22, File "NSF, 1976: General, April–August."

9. Michael H. Hunt, *The American Ascendancy: How the United States Gained and Wielded Global Dominance* (Chapel Hill: University of North Carolina Press, 2007), 225.

10. *Antarctica: Exploring the Frozen Continent*, An Encyclopaedia Britannica Educational Corporation Presentation in Association with ABC Owned Television Stations, 1979.

11. Michael Adas, *Dominance by Design: Technological Imperatives and America's Civilizing Mission* (Cambridge, MA: Harvard University Press, 2006), 281–336.

12. Walter LaFeber, *America, Russia, and the Cold War, 1945–2006* (Boston, MA: McGraw Hill, 2008), 276.

13. James T. Paterson, *Grand Expectations: The United States, 1945–1974* (New York: Oxford University Press, 1996), 451; "Man of the Year, Twenty Five and Under," *Time* (January 6, 1967), 18.

14. Michael A. Bernstein, "Understanding American Economic Decline: The Contours of the Late-Twentieth-Century Experience," in Michael A. Bernstein and David Adler, eds., *Understanding American Economic Decline* (Cambridge: Cambridge University Press, 1995), 3–33.

15. Commission on Critical Choices for Americans, *Vital Resources: Reports on Energy, Food, & Raw Materials* (Lexington, MA: Lexington Books, 1977), xxiii, xxx.

16. James T. Patterson, *Grand Expectations: The United States, 1945–1974* (New York: Oxford University Press, 1996), 317.

17. Commission on Critical Choices for Americans, *Vital Resources*, 80.

18. Barry Commoner, *The Closing Circle: Nature, Man & Technology* (New York: Bantam Books, 1971), 137.

19. Donella H. Meadows, et al., *The Limits to Growth: A Report for the Club of Rome's Project on the Predicament of Mankind* (New York: Universe Books, 1972), 23.

20. National Science Board Report, *Science and the Challenges Ahead* (Washington, DC: US Government Printing Office, 1974), 19.For government assessments of nonrenewable energy resources, also see: U.S. Department of Commerce, *United States Energy through the Year 2000* (Washington, DC: U.S. Government Printing Office, 1972); *The Nation's Energy Future: A Report to the President of the United States* (Washington, DC: U.S. Government Printing Office, 1973).

21. "Antarctic Resources: Report from the Meeting of Experts at the Fridtjof Nansen Foundation," in Senate Subcommittee on Oceans and International Environment, *U.S. Antarctic Policy*, 94th cong., 1st sess., 15 May 1975, 73.

22. H. Guyford Stever, "Director's Statement," *NSF Annual Report for 1975* (Washington, DC: National Science Foundation, 1975), vii.

23. Gerald O. Barney, Study Director, *The Global 2000 Report to the President of the U.S., Entering the 21st Century, Volume 1: The Summary Report* (New York: Pergamon Press, 1980), 1.

24. Herman Kahn et al., *The Next 200 Years: A Scenario for America and the World* (New York: William Morrow, 1976), 3.

25. Statement of the vice president at the White House Science Advisory Conference, June 6, 1975. Ford Library, Glenn R. Schleede Files, Box 38, file "Sci and Tech Policy, 1975: Vice Presidential Statements."

26. Heilbroner, *An Inquiry Into the Human Prospect*, 21.

27. Dr. Wernher von Braun, Deputy Associate Administrator, NASA, speech at the National Aviation Club, May 27, 1971. National Air and Space Museum Archives, OS-170884-54.

28. Banowsky, "Whatever Happened to the American Dream?," January 31, 1976; I. I. Rabi, "Government, Science and Technology, Twilight of the Gods," address before the International Atomic Energy Agency, Vienna, June 25, 1979. Library of Congress Archives, Papers of I. I. Rabi, Box 64.

29. The results of these polls were published in: *Science Indicators 1972: Report of the National Science Board, 1973* (Washington, DC: NSF, 1973).

30. Scholarly work on an apparent public "revolt" against science include: Gerald James Holton and William A Blanpied, eds., *Science and Its Public: The Changing Relationship* (Boston, MA: D. Reidel, 1976); Steven Louis Goldman, "Present Strains in the Relations between Science, Technology, and Society," *Science, Technology, and Human Values*, v. 4, n. 27 (Spring 1979), 44–51; Daniel Yankelovich, "Changing Public Attitudes to Science and the Quality of Life: Edited Excerpts from a Seminar," *Science, Technology, and Human Values*, v. 7, n. 39 (Spring 1982), 23–29.

31. Thomas Hughes, *American Genesis: A Century of Invention and Technological Enthusiasm, 1870–1970* (New York: Viking, 1989), 443–472.

32. On popular media treatment of science and technology during the 1960s and 1970s, see: Philip Hills and Michael Shallis, "Scientists and Their Images," *New Scientist* (August 28, 1975), 471–475; George Basalla, "Pop Science: The Depiction of Science in Popular Culture," Gerald James Holton and William A Blanpied, eds., *Science and Its Public: The Changing Relationship* (Boston, MA: D. Reidel, 1976); Roslynn Haynes, *From Faust to Strangelove: Representations of the Scientist in Western Literature* (Baltimore, MD: Johns Hopkins University Press), 1994.Scholarly work on press coverage of American science include: Dorothy Nelkin, *Selling Science: How the Press Covers Science and Technology* (New York: W. H. Freeman, 1987); John C. Burnham, *How Superstition Won and Science Lost: Popularizing Science and Health in the United States* (New Brunswick, NJ: Rutgers University Press, 1987.

33. Alan Waterman, "Science and International Affairs," September 14, 1959.

34. Thomas S. Kuhn, *The Structure of Scientific Revolutions* (Chicago: University of Chicago Press, 1962); Michael Mulkay, *Science and*

the Sociology of Knowledge (London: George Allen & Unwin, 1979); Jerry Gaston, "Sociology of Science and Technology," Paul T. Durbin et al., eds. *The Culture of Science, Technology, and Medicine* (New York: Free Press, 1980).

35. Aant Elzinga and Andrew Jamison, "Changing Policy Agendas in Science and Technology," in Sheila Jasanoff et al. eds., *Handbook of Science and Technology Studies* (London: Sage, 1995), 572–597.

36. Richard S. Lewis, "Antarctic Research and the Relevance of Science," in Richard S. Lewis and Philip M. Smith, eds., *Frozen Future: A Prophetic Report from Antarctica* (New York: Quadrangle Books, 1973), 2.

37. Leonard A. Redecke, NSF Deputy Director, Division of Grants and Contracts, diary note, "Comments of Dr. John Holmfeld, Science Policy Consultant to the Committee on Science and Technology, U.S. House of Representatives," April 5, 1975. National Archives, RG 307, NSF, E-121 Reading File, 1976 Records of the Head, Division of Polar Programs, Box 2.

38. Philip Smith note for Glenn Schleede and Dennis Barnes, September 2, 1976. Ford Library, Glenn R. Schleede Files, Box 39, File "Science and Technology Policy, 1976, General (1)."

39. Patterson, *Grand Expectations*, 790.

40. President's Commission for a National Agenda for the Eighties, *Science and Technology: Promises and Dangers in the Eighties* (Washington, DC: U.S. Government Printing Office, 1980), 13–14.

41. Gary Gerstle, *American Crucible: Race and Nation in the Twentieth Century* (Princeton, NJ: Princeton University Press, 2001), 4.

42. Lizabeth Cohen, *A Consumer's Republic: The Politics of Mass Consumption in Postwar America* (New York: Vintage Books, 2003), 292–344.

43. David Boyer, "Year of Discovery Opens in Antarctica," *National Geographic* (September 1957), 342; "Science in Antarctica," *Science* (January 26, 1968), 409.

44. Roger Tory Peterson, "Render the Penguins, Butcher the Seals: The Antarctic's Bloody Past May Foretell Its Future," *Audubon* (March 1973), 92.

45. Margaret Rossiter, *Women Scientists in America: Before Affirmative Action, 1940–1972* (Baltimore, MD: Johns Hopkins University Press, 1995), 377–378.

46. Tom Englehardt, *The End of Victory Culture: Cold War America and the Disillusioning of a Generation* (New York: Basic Books, 1995), 3–15.

47. Richard Slotkin, *Gunfighter Nation: The Myth of the Frontier in Twentieth-Century America* (New York: Atheneum, 1992), 624–633.

48. White House Press Release, December 12, 1974. Ford Library, Glenn R. Schleede Files, Box 8, file "Council on Environmental Quality 1974: Annual Report (1)."

49. Ezrahi, "The Political Resources of American Science," 124.
50. Victor Danilov, *America's Science Museums* (New York: Greenwood Press, 1990).
51. Metro Goldwyn Mayer, *The Secret Land*, 1948; Rear Admiral Richard Byrd, "All Eyes on the South Pole," *Reader's Digest* (December 1955), 117.
52. Oscar Handlin, ed., *American Principles and Issues: The National Purpose* (New York: Holt, Rinehart and Winston, 1961), v.
53. *Antarctica: The Coldest Continent* (Marble-Gibson Production and McDonnell-Douglas Corporation, 1968).
54. Ernest E. Angino "Burial of High-Level and Long-Lived Radioactive Wastes in Antarctica: A Reappraisal," *Antarctic Journal of the United States* (October 1979): 239–240; E. J. Zeller et al., "Putting Radioactive Wastes on Ice," *Bulletin of the Atomic Scientists* (January 1973), 4–9, 50–52; "Radioactive Wastes on Ice: Further Discussion," *Science and Public Affairs* (April 1973), 2–3, 53–56.
55. "Iceberg-Towing Plan Gets Legislature's Nod," *The Sacramento Bee* (April 1, 1978); Youssef M. Ibrahim, "Saudi Will Deliver Icebergs: At a Price," *New York Times* (April 15, 1978), A29; Information on a 1978 Iowa University conference and the related initiatives in the California legislature and in Congress, is in: National Archives, RG 307 NSF, E-122 Miscellaneous Records of the Director, 1978–79, Box 1, File "Correspondence 1978."
56. Walter Sullivan, "Why Antarctica Is Being Explored," *New York Times* (March 6, 1960), 20; History and Research Division, *Introduction to Antarctica* (Washington, DC: U.S. Naval Support Force, 1969), 48.
57. Samuel Matthews, "Antarctica's Nearer Side," *National Geographic* (November 1971), 654.
58. *Antarctica* (Washington, DC: National Science Foundation, 1974).
59. Samuel Matthews, "This Changing Earth," *National Geographic* (January 1973), 7; George Doumani, *The Frigid Mistress: Life and Exploration in Antarctica* (Baltimore, MD: Noble House, 1999).
60. N. A. Wright and P. L. Williams, *Mineral Resources of Antarctica*, Geological Survey Circular 705, Department of Interior, 1974.
61. Central Intelligence Agency, confidential report *Soviet Interests in Antarctic Mineral Resources*, September 1973, 4.
62. Deborah Shapley, *The Seventh Continent: Antarctica in a Resource Age* (Washington, DC: Resources for the Future, 1985).
63. Walter Sullivan, "Introduction," in *Frozen Future*, viii; "Antarctic Oil Is Estimated as Enormous," *Washington Post* (March 1, 1975), A6; Deborah Shapley, "Antarctica: World Hunger for Oil Spurs Security Council Review," *Science* (May 14, 1974): 777; Peter Gwynne, "Antarctica's Icy Assets," *Newsweek* (October 3, 1977), 93.
64. Patsy T. Mink, "Oceans: Antarctic Resource and Environmental Concerns," *Department of State Bulletin* (April 1978), 51.

65. *Polar Energy Resources Potential, Report Prepared for the House Subcommittee on Energy Research, Development and Demonstration* (Washington, DC: U.S. Government Printing Office, September 1976), xvii.On USGS estimates of Antarctic oil and gas and the concomitant interest of oil companies in Antarctic resources, see: NSF and USGS, John C. Behrendt ed., *Petroleum and Mineral Resources of Antarctica* (Alexandria, VA: US Geological Survey, 1983); Office of Technology Assessment, *Polar Prospects: A Minerals Treaty for Antarctica* (Washington, DC: US Government Printing Office, September 1989); Evan Luard, "Who Owns the Antarctic?," *Foreign Affairs* (Summer 1984), 1175–1193.

66. "Mineral Resource Exploitation in Antarctica: Outline of International Legal and Political Factors," September 7, 1972. National Archives, RG 307 NSF, E-64 Central Subject Files, 1969–1975, Box 4, file "Antarctic Policy Group."

67. United States Antarctic Program Study Report for Office of Management and Budget, May 1, 1972, vi. National Archives, RG 307, NSF, E-68, Miscellaneous Documents, 1969–75.

68. Robert Cushman Murphy, "Let's Make Antarctica an International Park," *Science Digest* (September 1962), 82.

69. Texts of these two measures are in: *Handbook of the Antarctic Treaty System*, 8th ed. (Washington, DC: U.S. Department of State, April 1994), 2048–2053, 156–161.

70. Leigh H. Fredrickson, "Environmental Awareness at Hallett Station," *Antarctic Journal of the United States*, v. 6, n. 3 (May–June 1971), 57.

71. Roderick Nash, *Wilderness and the American Mind* (New Haven, CT: Yale University Press, 1982), 254.

72. Heilbroner, *An Inquiry Into the Human Prospect*, 76; Samuel Hays, *Beauty, Health, and Permanence: Environmental Politics in the United States, 1955–1985* (Cambridge: Cambridge University Press, 1987).

73. National Science Foundation, *Nineteenth Annual Report, 1969* (Washington, DC: US Government Printing House, 1970), Foreword.

74. Donald Worster, *Nature's Economy: The Roots of Ecology* (San Francisco: Sierra Club, 1977), 318.

75. John McCormick, *Reclaiming Paradise: The Global Environmental Movement* (Bloomington: Indiana University Press, 1989), 133, 104–105.

76. "President's Commission for the Observance of the 25th Anniversary of the United Nations," *Department of State Bulletin* (August 2, 1971), 132.

77. Laurence M. Gould, "Antarctica: The World's Greatest Laboratory," *The American Scholar* (Summer 1971), 408.

78. Raymond Dasmann, "Conservation in the Antarctic," *Antarctic Journal of the United States*, v. 3, n. 1 (January–February 1968), 1.

79. Bruce C. Parker, "The Case for Conservation in Antarctica," *Antarctic Journal of the United States*, v. 6, n. 3 (May–June 1971), 50.

80. CSAGI, *Bulletin d'Information*, No. 4 (London: IUGG Newsletter No. 9, 1955), 54–55; D. W. H. Walton, ed., *Antarctic Science* (Cambridge: Cambridge University Press, 1987), 33.

81. Ronald Fraser, *Once Round the Sun: The Story of the International Geophysical Year* (New York: Macmillan, 1961), 119.

82. Spencer Weart, *The Discovery of Global Warming* (Cambridge, MA: Harvard University Press, 2003).

83. Popular works in the 1970s anticipating global climate change, particularly global cooling, include: Nigel Calder, *The Weather Machine* (New York: Viking, 1974); Lowell Ponte, *The Cooling* (Englewood Cliffs, NJ: Prentice-Hall, 1976); John Gribben, *Forecasts, Famines and Freezes: Climate and Man's Future* (New York: Walker, 1976); D.S. Halacy, *Ice or Fire? Can We Survive Climate Change?* (New York: Barnes & Noble Books, 1978).

84. Samuel Matthews, "What's Happening to Our Climate?," *National Geographic* (November 1976), 579.

85. Dr. Owen to Dr. Stever, July 5, 1973. National Archives, RG 307, NSF, E-69, Memorandums, 1971–74, Records of the Head OPP, Box 1, file "Director, 1973."

86. H. Guyford Stever memo to President Nixon. National Archives, RG 307 NSF, E-74, Program Documents, 1969–75, Office of Polar Programs, Box 1 file "A National Program for the Study of Climate Fluctuations and Their Impact on Human Affairs, 1973–1974."

87. "President Nixon Announces Review of U.S. Policy for Antarctica," *Department of State Bulletin*, November 9, 1970, 572.

88. Secretary Rogers, "Statement by President Nixon," *Department of State Bulletin* (July 19, 1971), 82.

89. Lewis, "Antarctic Research and the Relevance of Science," 2.

90. *Antarctica* (Washington, DC: National Science Foundation, 1974).

91. "President Nixon Announces Review of U.S. Policy for Antarctica," *Department of State Bulletin* (November 9, 1970), 573; "President's Commission for the Observance of the 25th Anniversary of the United Nations," *Department of State Bulletin* (August 2, 1971), 132.

92. "Recommendations of the Second World Conference on National Parks," *Supplement to IUCN Bulletin 3*, November 1982, 2.

93. An NSF Office of Polar Programs report regretted that it was unlikely to get money from Congress needed to build a new airstrip in Antarctica owing to the "fiscal impracticality of a three year non defense, non welfare, non pork barrel Federal capital expenditure is a pure pipe dream." See: "McMurdo Station Familiarization for C.N. Chamberlain Trip Report," 1973. National Archives, RG 307, NSF, E-120 Central Subject Files, 1976–87, Box 3, file "McMurdo Station (General)."

94. Central Intelligence Agency, *Soviet Interests in Antarctic Mineral Resources*, 9.

95. *Handbook of the Antarctic Treaty System*, 191.

96. Ibid., 194–195.

97. Ibid., 172.

98. Ibid., 178–188.

99. "Parties to Antarctic Treaty Meet in London," *Department of State Bulletin* (November 21, 1977), 739.

100. Patsy Mink, *Exploitation of Antarctic Resources*, 95th cong., 2nd sess., February 6, 1978, 18.

101. Testimony of Leonard C. Meeker, *Exploitation of Antarctic Resources*. Senate Subcommittee on Arms Control, Oceans, and International Environment, 95th Cong., 2nd sess., February 6, 1978, 93.

102. "Trip to the Bottom of the World," *Time* (January 5, 1976), 58; "Heating Up: Global Race for Antarctic's Riches," *U.S. News & World Report* (February 28, 1977), 65; Gwynne, "Antarctica's Icy Assets," 92.

103. Testimony of James Barnes, Senate Committee on Commerce, Science, and Transportation, *Antarctic Living Marine Resources Negotiations*, 95th cong., 2nd sess., June 14, 1978, 32.

104. "1979 Letter to President Carter," and "1980 Letter to President Carter" in James N. Barnes, *Let's Save Antarctica* (Richmond, Australia: Greenhouse, 1982), 76–77.

105. SCAR Bulletin, "Antarctic Resources-Effects of Mineral Exploration," *Polar Record* (September 1977), 386.

106. Christopher C. Joyner, "The Antarctic Minerals Negotiating Process," *The American Journal of International Law*, v. 81 (1987), 891.

107. *Handbook of the Antarctic Treaty System*, 198.

108. "International Union for Conservation of Nature and Natural Resources: 1981 General Assembly Resolution," in Edwin Mickleburgh, ed., *Beyond the Frozen Sea: Visions of Antarctica* (New York: St. Martin's Press, 1987), 219; "Resolution of Non Government Organizations Concerning Antarctica and the Southern Ocean," in James N. Barnes, *Let's Save Antarctica* (Richmond, Australia: Greenhouse, 1982), 64.

109. Barnes, *Let's Save Antarctica*, 26.

110. Testimony of James Barnes, Senate Subcommittee on Science, Technology, and Space, *Antarctica*, 98th cong., 2nd sess., September 24, 1984, 67–68.

111. Barnes, *Let's Save Antarctica*, 36.

112. "President's Memorandum Regarding Antarctica," *Facts about the U.S. Antarctic Program* (Washington, DC: NSF Division of Polar Programs, 1988), 19; "Department of State Issues an Environmental Impact Statement on Mineral Resources," *Antarctic Journal of the United States* v. 17, n. 4 (December 1982), 6.

113. R. Tucker Scully, Senate Subcommittee on Science, Technology, and Space, *Antarctica*, 98th cong., 2nd sess., September 24, 1984, 2, 6–7.

114. Keith Suter, *Antarctica: Private Property or Public Heritage* (New Jersey: Zed Books, 1991), 75.

115. Ibid., 82.

116. R. Tucker Scully, Senate Subcommittee on Science, Technology, and Space, *Antarctica*, 98th cong., 2nd sess., September 24, 1984, 7.

117. John D. Negroponte, *The Success of the Antarctic Treaty* (Washington, DC: U.S. Department of State, Bureau of Public Affairs, April 1987), 1.

118. Lee Kimball, "The Role of Non-Governmental Organizations in Antarctic Affairs," in Christopher C. Joyner and Sudhir K. Chopra, eds., *The Antarctic Legal Regime* (Boston, MA: Martinus Nijhoff, 1988), 39.

119. *Antarctic Treaty System: An Assessment. Proceedings of a Workshop Held at Beardmore South Field Camp, Antarctica, January 7–13, 1985* (Washington, DC: National Academy Press, 1986).With support from several major American philanthropic foundations, the National Science Foundation and the US Polar Research Board organized the workshop and airlifted 57 people from 25 countries to the remote location.

120. Suter, *Antarctica*, 85.

121. J. C. Farman et al., "Large Losses of Total Ozone in Antarctica Reveal CLOx/Nox Interaction," *Nature* (May 16, 1985).

122. Sharon Roan, *Ozone Crisis: The 15-Year Evolution of a Sudden Global Emergency* (New York: John Wiley, 1989).

123. Ibid., v.

124. Richard E. Benedick, "Ozone Depletion," *Buzzworm* (January/ February 1991), 38.

125. Stephen H. Schneider, "Global Warming," *Buzzworm* (January/ February 1991), 39.

126. Roan, *Ozone Crisis*; *Time* (January 2, 1989), 66.

127. Cover page, *National Geographic* (December 1988).

128. *Time* (January 2, 1989), 2.

129. Richard S. Stolarski, "The Antarctic Ozone Hole," *Scientific American* (January 1988), 30, 36.

130. Allan Hall, "The World's Frozen Clean Room, *Business Week* (January 22, 1990), 73.

131. *Focus on Antarctica* (Science News Digest, Hearst Productions, 1960); *Antarctica* (Omnimax Films, 1991).

132. "Nightline in Antarctica," *ABC News Nightline* (November 24, 1992).

133. James Gleick, "A Huge Chunk of Antarctica is Heading Out to Sea, at a Glacial Pace," *The New York Times* (February 9, 1988), C4.

134. Traveling Exhibit, *Antarctica*, Boston Museum of Science, 1992.

135. "Antarctica: Polar Wilderness in Peril," *National Geographic World* (February 1991), 3. In "The World's Frozen Clean Room," *Business Week* (January 22, 1990), 73; Allan Hall similarly wrote that "[m] any scientists believe that the Antarctic atmosphere provides an early warning system for global warming."

136. Walton, *Antarctic Science*, 220.

137. Richard Laws, "Science as an Antarctic Resource," in Graham Cook, ed., *The Future of Antarctica: Exploitation versus Preservation* (New York: Manchester University Press, 1990), 10.

138. John May, *The Greenpeace Book of Antarctica: A New View of the Seventh Continent* (New York: Doubleday, 1989), 2.

139. "Engineering Feat or Blight?," *Nature*, v. 350 (March 28, 1991), 301; M. Lynne Corn et al., *Environmental Effects of Recent Activities in Antarctica*, Congressional Research Service, Library of Congress, June 15, 1988, CRS-33–36.

140. May, *The Greenpeace Book of Antarctica*, 3.

141. Testimony of Greenpeace, Senate Subcommittee on Science, Technology, and Space, *Protecting Antarctica's Environment*, 101st cong., 1st sess., September 8, 1989, 179.

142. Kelly Rigg, "Environmentalists' Perspectives on the Protection of Antarctica," *The Future of Antarctica*, 68–69.

143. Senate Subcommittee on Science, Technology, and Space, *Arctic and Antarctic Ozone Depletion*, 101st cong., 1st sess., February 23, 1989, 3.

144. House Subcommittee on Human Rights and International Organizations, *Preserving Antarctica's Ecosystem*, 101st cong., 2nd sess., May 2 and July 19, 1990, 1.

145. "President's Message to Antarctic Scientists," *Antarctic Journal of the United States* v. 23, n. 3 (September 1988), 20.

146. In its *1983 Annual Report*, the NSF described in very general terms USAP research and measures to protect the local environment. In its *1987 Annual Report*, however, the NSF focused on ozone deterioration and USAP research on this global environmental crisis. And in its *1988 Annual Report*, the NSF devoted a whole section to global environmental change and pointed out the many ways the agency supported American scientists who studied this important issue.

147. *The United States Antarctic Program* (Washington, DC: NSF Division of Polar Programs, 1990), 4.

148. Senate Committee on Foreign Relations, *Antarctic Legislation*, 101st cong., 2nd sess., July 27, 1990, 17.

149. Peter Beck, "Antarctica Enters the 1990s: An Overview," *Applied Geography*, v. 10, n. 4 (1990), 260. Also see: James E. Miekle and Marjorie Ann Browne, *Antarctic Mineral Resources: Diplomacy and Dilemma*, Congressional Research Service Issue Brief (Library of Congress, August 29, 1991).

150. Paul Larmer, "The Great White Heap?," *Sierra* (March/April 1990), 27.

151. Michael D. Lemonick, "Antarctica," *Time* (January 15, 1990), 56, 62.

152. Isaac Asimov, *How Did We Find Out about Antarctica* (New York: Walker, 1979), 62.

153. "Antarctica: Polar Wilderness in Peril," *National Geographic World* (February 1991), 3. Environmental protection was the primary message in the children's books: Helen Coucher, *Antarctica* (New York: Farrar, Straus and Giroux, 1990); Ulco Glimmerveen, *A Tale of Antarctica* (New York: Scholastic, 1989).

154. House Committee on Merchant Marine and Fisheries, *Antarctic Briefing with Jacques-Yves Cousteau*, 101st cong., 2nd sess., May 2, 1990, 4.

155. "Protection of the Antarctic Environment," *GIST* (Washington, DC: State Department Bureau of Public Affairs, May 1990).

156. Suter, *Antarctica*, 59.

157. Whereas the May 1990 copy of *GIST*, the State Department's "quick reference aid on US foreign policy," strongly advocated CRAMRA, its October 8, 1992 *GIST* on the "Protection of the Antarctic Environment" did not even mention CRAMRA as it detailed America's past positions on the Antarctic environment and its strong endorsement of the current 50-year moratorium on Antarctic mineral development.

158. *Protocol on Environmental Protection to the Antarctic Treaty: Message from the President of the United States* (Washington, DC: U.S. Government Printing House, 1992), iv.

159. Stanley Johnson, *Antarctica: The Last Great Wilderness* (London: Weidenfeld and Nicolson, 1985), 184.

160. "Science and the Environment," *Antarctic Journal of the United States*, v. 24, n. 4 (December 1989), 13; Adrian Howkins, "Melting Empires?: Climate Change and Politics in Antarctica since the International Geophysical Year," *Osiris*, v. 26 (2011), 180–197.

161. "Expedition Antarctica" (Mountain Travel-Sobek Company, 1993).

162. David E. Nye, *American Technological Sublime* (Cambridge, MA: MIT Press, 1996).

163. *Safety in Antarctica: Report of the U.S. Antarctic Program Safety Review Panel* (Washington, DC: National Science Foundation, 1988), I-6.

164. Daniel Grotta and Sally Grotta, "Antarctica: Whose Continent Is It Anyway?," *Popular Science* (January 1992), 63.

165. Edwin Mickleburgh, *Beyond the Frozen Sea: Visions of Antarctica* (New York: St. Martin's Press, 1987), 4.

166. *Safety in Antarctica*, I-6.

167. "Fuel Spill Clean Up in Antarctica," *Antarctic Journal of the United States*, v. 25, n. 4 (December 1990), 3–10.

168. Caricatures of the truck "Greenpiece" appear in the *Antarctic Sun Times*—a newsletter that was published at McMurdo station for US Antarctic personnel. Pamphlet-collection of the State Historical Society of Wisconsin.

169. Jacques Cousteau, "Life Insurance," *Calypso Log* (April 1990), 3.

170. Barry Lopez, "Informed by Indifference," *Harper's Magazine* (May 1988), 67; Sir Peter Scott, "Foreword," in *The Greenpeace Book of Antarctica*.

171. *Greenpeace Book of Antarctica*, "Introduction."

172. Herman Melville, *Moby Dick* (New York: Signet Classics, 1961), 196.

4 The Grip of the Space Frontier

1. *Time* (January 6, 1967), 18.

2. "Men of the Year," *Time* (January 3, 1969), 4.

3. Richard Nixon, "Statement about the Future of the United States Space Program," March 7, 1970.

4. SNCC letter to Representative Adam Clayton Powell, January 13, 1964. Lyndon B. Johnson Presidential Library.

5. "Destination 'Moon,'" *Ebony*, v. 13 (July 1958), 120–124; "Space Doctor for the Astronauts," *Ebony*, v. 17 (April 1962), 35–39; "Angel of Mercy to the Astronauts," *Ebony*, v. 21 (June 1966), 49–52.

6. Robert McAfee Brown, "Moon Shot and Afterthoughts," *Christianity and Crisis*, v. XXIX, n. 15 (September 15, 1969).

7. Steven Morris, "How Blacks View Mankind's 'Giant Step,'" *Ebony* v. 25 (September 1970), 33–42.

8. Stan Scott memo to Ken Clawson, "NASA's Image," February 19, 1974. National Archives, Nixon Presidential Material Project, White House Central Files, FG 164 National Aeronautics and Space Administration, Box 2, file "[EX] FG 164 NASA 1/1/73"; Kim McQuaid, "'Racism, Sexism, and Space Ventures': Civil Rights at NASA in the Nixon Era and Beyond," in Steven J. Dick and Roger D. Launius, eds., *Societal Impact of Spaceflight* (Washington, DC: NASA, 2007), 421–450.

9. Fred Malek (Deputy Director OMB) letter to James Fletcher (NASA Administrator), February 4, 1974, Gerald R. Ford Library, Glenn R. Scheede Papers, Box 20, File NASA, 1974: Budget for FY 1976 (2); Henry Capote (Hispanic Program Manager, NASA) letter to Gil Colon (Deputy Assistant to the President for Hispanic Affairs) November 27, 1979, Jimmy Carter Library, Staff Offices, Hispanic Affairs-Colon, Box 9, file NASA 1979–80.

10. "All Female Space Crew Is Feasible, Scientists Agree," *The Houston Post* (March 6, 1978).

11. Joan McCullough, "The 13 Who Were Left Behind," *Ms.* (September 1973), 45.

12. H. Malone, "Ultimate Phallic Journey," *Fifth Estate*, v. 4, n. 7 (August 20, 1969), 4; Berkeley Gazette, "Male Chauvinism on the Moon," *It Ain't Me Babe*, v. 1, n. 2 (January 29, 1970), 10.

13. Ena Nauton, "A Woman in Outer Space? Good Idea—But Still a Joke," *The Miami Herald* (May 21, 1972), 17; Jack Anderson, "Would-Be Astronauts: Legion of Angry Women," *Parade* (November 19, 1967).

14. Michael Smith, "Selling the Moon: The U.S. Manned Space Program and the Triumph of Commodity Scientism" in T. J. Jackson Lears and Richard Wightman Fox, eds., *The Culture of Consumption: Critical Essays in American History, 1880–1980* (New York: Pantheon, 1983), 205.

15. Dr. Wernher von Braun, Speech at the National Aviation Club, 27 May 1971. Smithsonian National Air and Space Museum (NASM) Archives, OS-170884-54.

16. James T. Patterson, *Grand Expectations: The United States, 1945–1974* (New York: Oxford University Press, 1996), 317.

17. Roger Launius, "A Waning of Technocratic Faith: NASA and the Politics of the Space Shuttle Decision," *Journal of the British Interplanetary Society*, v. 49 (1996), 49–58.

18. Loudon Wainwright, "The Dawn of the Day Man Left His Planetary Circle," *Life* special edition, "To the Moon and Back," (New York: Time, 1969).

19. Ronald Weber, *Seeing Earth: Literary Responses to Space Exploration* (Athens: Ohio University Press, 1985), 8.

20. "Men of the Year," 4.

21. Richard Nixon, "Telephone Conversation with the Apollo 11 Astronauts on the Moon," July 20, 1969.

22. "Statement by George Mueller," July 24, 1969. Ford Library, H. Guyford Stever Papers, 1930–90, Box 66, file "NASA (1967–70)."

23. Statement by Vice President Spiro Agnew, Chairman of the National Aeronautics and Space Council, May 21, 1969. NASM, OS-170884-50.

24. Homer E. Newell, "Review and Assessment of the U.S. Space Program," May 7–8, 1971 presentation at MIT. National Archives, RG 255 NASA, Dr. Homer Newell, Miscellaneous, Box 108, unnamed file.

25. Dr. Krafft Ehricke, "The 'Absolute Necessity' of Space Exploitation," *Space World* (June 1970), 38.

26. Letter to President Nixon signed by 27 children, May 20, 1969. National Archives, Nixon Presidential Material Project, White House Central Files, FG 164 National Aeronautics and Space Administration, Box 2, file "[Gen] NASA 3/31/70."

27. Brian Duff to William L. Green, "Fiscal Year 1969 Planning," September 1, 1967. NASA History Office (NHO), Office of Public Affairs Box, file "Chronological File, 1963–1968."

28. NASA Special Announcement, "Apollo 11 Tour of 50 State Capitals," December 1970. NHO, Office of Public Affairs Box, file "Public Affairs Exhibits." And NASA News Release, "40-Plus Million See Moon Rocks," January 20, 1971, NASA Release No.: 71–8.

29. Julian Scheer memo to NASA Administrator, September 10, 1969. National Archives, Nixon Presidential Material Project, White House Central Files, Subject Files, Outer Space, Box 1, file "{EX} OS Outer Space [1969–1970]."

30. Herbert E. Krugman, "Public Attitudes toward the Apollo Space Program, 1965–1975," *Journal of Communication*, v. 27 (Autumn 1977), 87.

31. Peter Flanigan Memo for the President, December 6, 1969. John M. Logsdon et al. eds., *Exploring the Unknown: Selected Documents in the History of the U.S. Civil Space Program, Vol. I: Organizing for Exploration* (Washington, DC: NASA, 1995), 546.

32. President's Space Advisory Committee, *The Space Program in the Post-Apollo Period* (Washington, DC: The White House, February 1967); President's Space Task Force, *The Post-Apollo Space Program: Directions for the Future* (September 1969).

33. T. O. Paine memo for the President, "NASA Activities," February 4, 1969. National Archives, Nixon Presidential Material Project, White House Central Files, Subject Files, Outer Space, Box 4, file "[EX] OS 3 Space Flight 3/31/69."

34. Secret Report "Space Goals after the Lunar Landing," October 1966. National Archives, RG 59 Department of State, 1964–66 Central Files, Box 3145, file "SP1 Gen. Policy.Plans.Coordination. 1/1/64"; Confidential "Space Goals and Foreign Policy, Priorities," May 10, 1967. National Archives, RG 59 Department of State, 1964–66 Central Files, Box 2870, file "SP 10 US 11/1/67."

35. Dr. Wernher von Braun, "Why Is Space Exploration Vital to Man's Future," *Space World* (September 1969), 31–33.

36. Richard M. Nixon, presidential candidate press release, "The Scientific Revolution," September 8, 1960, National Archives, RG 307 National Science Foundation, Entry-39 Office of Antarctic Programs Central Files, Box 124, 110.1, "The White House Press Release."

37. Donella H. Meadows et al., *The Limits to Growth: A Report for the Club of Rome's Project on the Predicament of Mankind* (New York: Universe Books, 1972), 23.

38. Willis Shapley interview, in Martin Collins, ed., *Oral History on Space, Science, and Technology* (Washington, DC: National Air and Space Museum, Smithsonian Institution, 1993), 9.

39. Robert P. Mayo (Director BOB) memo for the President, "Space Task Group Report, September 25, 1969," *Exploring the Unknown*, 1995, 545; Nixon, "Statement about the Future of the United States Space Program."

40. Daniel P. Moynihan to General George A. Lincoln, August 13, 1969. National Archives, Nixon White House Central Files, FG 164 National Aeronautics and Space Administration, box 1, file "[EX] NASA 8/31/69."

41. Bill Safire to H. R. Haldeman, May 17, 1971. National Archives, Nixon White House Central Files, FG 164 National Aeronautics and Space Administration, Box 2, file "[EX] FG 164 NASA 1/1/71."

42. *Aeronautics and Space Report of the President, Fiscal Year 1994 Activities* (Washington, DC: NASA, 1995), 93; James C. Fletcher note for Stuart Eizenstat (Assistant to the President), February 16, 1977. Jimmy Carter Library, Domestic Policy Staff, Eizenstat Files, Box 238, file "N.A.S.A. [O/A 6465]."

43. Nixon, "Statement about the Future of the United States Space Program." *Public Papers of the Presidents: Richard Nixon, 1970* (Washington, DC: Office of the Federal Register, 1971), 250–251.

44. "The Sun and the Planets," *Washington Post* (March 9, 1970); "Space in the 1970s," *New York Times* (March 10, 1970).

45. Testimony of David S. Lewis, *The National Space Program: Present and Future, a Compilation of Papers*, House Committee on Science and Astronautics, 91st cong., 2nd sess., December 10, 1970, 63.

46. Dr. Thomas O. Paine, "What Lies Ahead in Space," Talk to the Economic Club of Detroit, September 14, 1970. NASM Archives, file OS 170889–52.

47. Office of the White House Press Secretary, "Press Conference of Dr. Thomas O. Paine, Administrator, National Aeronautics and Space Administration, March 7, 1970. NASM Archives, OS-170889–52.

48. Bruce L. R. Smith, *American Science Policy Since World War II* (Washington, DC: Brookings Institution, 1990), 86.

49. *Technology Utilization Program Report* (Washington, DC: NASA SP-5119, December 1973).

50. Karl G. Harr, "Space and Tomorrow's Society," *Spaceworld* (December 1970), 31.

51. Dr. Wernher von Braun, "Welcome and Introduction to Assembly," *Space for Mankind's Benefit: Proceedings of a Space Congress Held November 15–19, 1971* (Washington, DC: NASA, 1972), 13.

52. Robert Holtz, "The Ecological Problem," *Aviation Week & Space Technology* (April 12, 1971), 11.

53. NASA SP-230, *Ecological Surveys from Space* (Washington, DC: NASA, 1979), 4.

54. Statement of Charles W. Mathews, Associate Administrator for Applications, NASA, "Pollution Monitoring Program," FY 75 Congressional Testimony. NASM OE-008 030–10.

55. NASA New Releases, "NASA Anti-Pollution Experiment," August 9, 1970, No. 70–133; "Oil Slick Detection Development," March 19, 1971, No. 71–42; "Space Technology to Aid Smog Research," June 16, 1971, No. 71–108; "Remote Measurements of Pollution," March 14, 1972, No. 72–58; "Satellite Helps Protect Florida Everglades," September 18, 1974, No. 74–263; "Study of Freon Threat under Way at NASA," April 28, 1975, No. 75–124.

56. NASA, *Improving Our Environment* (Washington, DC: U.S. Government Printing Office, 1973); NASA Fact Sheets, *Earth Resources* (Washington, DC: U.S. Government Printing Office, 1973).

57. *NASA and Energy* (Washington, DC: NASA EP-121, 1974), 1.

58. Thomas Paine letter to the President, September 1, 1970. National Archives, Nixon White House Central Files, Subject Files, Outer Space, Box 7, file "[EX] OS 3 9/1/70–10/31/70."

59. Edward David memo for the President, December 31, 1970. National Archives, Nixon White House Central Files, Subject Files, Outer Space, Box 7, file "[EX] OS 3 11/1/70–12/31/70."

60. Ed Harper to Counselor Robert Finch, "Key Election Issue: Federally Caused Unemployment," September 23, 1970. National Archives, Nixon White House Central Files, FG 164 National Aeronautics and Space Administration, Box 1, file "[EX] FG 164 8/1/70–9/30/70"; John Ehrlichman memo to the President, "Moon Shot and Other NASA Cuts," May 27, 1971. National Archives, Nixon White House Central Files, Subject Files, Outer Space, Box 8, file "[EX] OS 3 5/1/71–5/31/71."

61. Dave Gergen Memo for Noel Koch, "Plaque for Apollo 17 Crew," November 1972. National Archives, Nixon White House Central Files, Subject Files, Outer Space, Box 9, file "[EX] OS 3 1/1/72 [1971–1972]."

62. "Up and Down for NASA," *Nature* (May 2, 1970).

63. "Annual Message to the Congress on the State of the Union," January 20, 1972. *Public Papers of the President, Richard M. Nixon, 1972* (Washington, DC: Office of the Federal Register, 1973), 71.

64. Ken Hechler, *The Endless Space Frontier: A History of the House Committee on Science and Astronautics, 1959–1978, AAS History Series, Vol. 4* (San Diego: AAS, 1982), 261; "Investment in the Future," *New York Times* (January 8, 1972).

65. "Space: A Report to Stockholder," *CBS Reports*, July 22, 1974. Library of Congress (LOC) Film and Television Archives.

66. Jesco von Puttkamer, "Space for Mankind's Benefit: A Space Congress for the Nonaerospace Public, Message from the Chairman," *Space for Mankind's Benefit*, 3; Ivan Rattinger, "The Potential Impact of the Space Shuttle on Space Benefits to Mankind," *Space for Mankind's Benefit*, 64, 65.

67. Senator Alan Cranston congressional testimony, reprinted as "Looking Out for Our Earth," *Spaceworld* (September 1973); ...*for the Benefit of All Mankind* (Philadelphia: General Electric, 1972).

68. Robert Jastrow and Homer Newell, "The Space Program and the National Interest," *Foreign Affairs* (April 1972), 535; "Measuring the Impact of NASA on the Nation's Economy," NASA Office of Special Studies, September 1990. NHO, file "Impact: Economic (1987–1990)"; James C. Fletcher letter to the Vice President, May 15, 1975. Ford Library. Glenn R. Schleede Files, Box 70, file "NASA, 1975: General, April–June." This data came in: Chase Econometric Associates, Inc., "The Economic Impact of NASA R&D Spending," prepared for NASA, April 1975.

69. Roger D. Launius, "A Western Mormon in Washington, D.C.: James C. Fletcher, NASA, and the Final Frontier," *Pacific Historical Review* v. 64, n. 2 (1995), 222; Clay Whitehead memo for Peter Flanigan, February 6, 1971. National Archives, FG 164 National Aeronautics and Space Administration, Box 2, file [EX] FG 164 N.A.S.A. 1/1/71.

70. Memo "The U.S. Space Program in the 1980s," December 10, 1971. National Archives, RG 255 NASA Records of Dr. Homer Newell, Miscellaneous, Box 113, file "Thrusts for the 1980s and Technology in 1970s"; Remarks by James Fletcher to the Salt Lake Rotary Club, June 6, 1972. NASM Archives, file OS-170889–56.

71. Glenn Schleede memo to Jim Cannon, "NASA's 1977 Budget," November 10, 1975. Ford Library, Glenn Schleede Files, Box 21, file "NASA, 1975: Budget."

72. *NASA and the "Now" Syndrome* (Washington, DC: NASA, 1975), 2.

73. Launius, "A Western Mormon in Washington, D.C.," 217–218.

74. Thomas Hughes, *American Genesis: A Century of Invention and Technological Enthusiasm, 1870–1970* (New York: Viking, 1989), 443–472.

75. "Reaching Beyond the Rational," *Time* (April 23, 1973), 95; President's Commission for a National Agenda for the Eighties, *Science and Technology: Promises and Dangers in the Eighties* (Washington, DC: U.S. Government Printing House, 1980), 14.

76. Gerald O. Barney, Study Director, *The Global 2000 Report to the President of the U.S., Entering the 21st Century, Volume 1: The Summary Report* (New York: Pergamon Press, 1980), 1; Jimmy Carter, "Voyager Spacecraft Statement by the President," July 29, 1977.

77. Jimmy Carter, "National Medal of Science Remarks at the Presentation Ceremony," January 14, 1980.

78. Herman Kahn et al., *The Next 200 Years: A Scenario for America and the World* (New York: William Morrow, 1976), 5.

79. Michael A. G. Michaud, *Reaching for the High Frontier: The American Pro-Space Movement, 1972–84* (New York: Praeger, 1986), 41.

80. "Harvest Moon," NASM, Office of the Director, Records, Box 306, folder "Committee for the Future"; Barbara Hubbard letter to Henry Kissinger, August 2, 1971. National Archives, Nixon White House Central Files, Subject Files, Outer Space, Box 11, file [GEN] OS3 Space Flight 1/1/71–9/30/71.

81. Gerard K. O'Neill, *The High Frontier: Human Colonies in Space* (Princeton, NJ: Space Studies Institute Press, 1989), 251; W. Patrick McCray, *The Visioneers: How a Group of Elite Scientists Pursued Space Colonies, Nanotechnologies, and a Limitless Future* (Princeton, NJ: Princeton University Press, 2012).

82. O'Neill, *The High Frontier*, 263; John C. Mankins, *The Case for Space Solar Power* (Houston, TX: Virginia Edition, 2014).

83. O'Neill, *The High Frontier*, 17.

84. Stewart Brand, ed., *Space Colonies* (New York: Penguin, 1977), 37.

85. T. A. Heppenheimer, *Colonies in Space* (Harrisburg: Stackpole Books, 1977).

86. J. Peter Vajik, *Doomsday Has Been Cancelled* (Culver City: Peace Press, 1978), xiii.

87. Walter Sullivan, "Proposal for Human Colonies in Space Is Hailed by Scientists as Feasible Now," *New York Times* (May 13, 1974), 1;

88. Isaac Asimov, "The Next Frontier?," *National Geographic* (July 1976), 76–89.

89. CBS, "Space Colonization," *60 Minutes*, October 9, 1977. LOC Television and Film Archives.

90. *Walt Disney World: A Pictorial Souvenir Featuring the Magic Kingdom and EPCOT Center* (Walt Disney Co., 1986), 30–31.

91. Paul T. Libassi, "Space to Grow," *The Sciences* (July/August 1974), 20.

92. Brian O'Leary, "Project Columbus," *Bulletin of the Atomic Scientists* (March 1977), 4–5; Michael J. Gaffey and Thomas B. McCord, "Mining Outer Space," *Technology Review* (June 1977); Michael Modell, "Sustaining Life in a Space Colony," *Technology Review* (June 1977).

93. "NASA's Lack of a 'Far Out' Spokesman," March 11, 1974; Assistant Administrator for Public Affairs memo to Associate Administrator for External Affairs, December 2, 1975. NHO, file "John P. Donnelly 000488."

94. David Ebbe, "Cities of the Sun," *Science Digest* (October 1982), 67; documents on NASA involvement in organizing the National Space Institute are in: NHO, file 005005.

95. Richard D. Johnson and Charles Holbrow, eds., *Space Settlements: A Design Study* (Washington, DC: NASA, 1977), 182; Aeronautics and Space Report #137, *Space Colonization*, NASA (July 1976). NASM Film Archives, FA 00464; James C. Fletcher letter to John O. Marsh, Jr. (Counselor to the President), August 6, 1976. Ford Library, John Marsh Files, Box 109, file "Fletcher, James C." Remarks by George Low at the Utah Air Force Association Bicentennial Program, April 23, 1976. NHO, file "Impact: Future (1929–1979)."

96. Office of Technology Assessment, *OTA Priorities, 1979* (Washington, DC: Congress of the United States, January 1981), 10.

97. *L5 Space Magazine*, "video producer Greg Barr, Tucson Community Cable Corporation" (December 1985).

98. Frank Press memo to the President, "Administration Space Policy," August 30, 1977. Jimmy Carter Library, WPS, Stern, Box 44, file "NASA: Space Policy."

99. Tom Englehardt, *The End of Victory Culture: Cold War America and the Disillusioning of a Generation* (Amherst: University of Massachusetts Press, 1995), 266–269.

100. Space Exploration Day, 1976, By the President of the United States, A Proclamation, July 19, 1976. Ford Library, Schleede, Box 21, File "NASA, 1976: General, July–October."

101. "NASA Recommendations," Nick MacNeil memo, Carter-Mondale Transition Planning Group, January 15, 1977. Jimmy Carter Library, Stern, Box 4, file "NASA."

102. "Space Settlements, Other Usage Proposals," NASA Information Sheet, 78–3.

103. NASA News, *Satellite Power Station Study* (Washington, DC: NASA, July 6, 1972); Aeronautics and Space Report #139, *Energy From Space*, NASA, September 1976. NASM Film Archives, FA 00465; Aeronautics and Space Report #147, *Space Solar Power*, NASA, May 5, 1977, NASM Film Archives, FA 00469; *Solar Power Satellite: Power from Space...a New Opportunity* (Washington, DC: NASA, November 13, 1978).

104. NASA's E. Z. Gray memo to agency Administrator and Deputy Administrator, "Satellite Solar Power Station," February 8, 1974. NHO, file "Solar Powered Satellites"; Testimony of Dr. Robert Frosch, *1979 NASA Authorization*, House Subcommittee on Space Science and Applications, 95th cong., 2nd sess., February 8, 9, 22, 23, 1978, 2588.

105. George Low (NASA Deputy Administrator) letter to Dr. George Solomon (Chairman NRC's Aeronautics and Space Engineering Board), May 15, 1975. NASA History Office, file "Solar Powered Satellites"; Robert Seamans letter to James Fletcher, September 30, 1975. Ford Library, Glenn R. Schleede Files, Box 21, file "NASA, 1975: Space Shuttle (1)"; Jim Wright letter to the President, March 29, 1977. NHO, file "Solar Powered Satellites."

106. Sam Dunham (Texas Realtor) letter to Frank Moore (White House Staff), August 23, 1977. Jimmy Carter Library, White House Central Files, Subject Files, Outer Space, Box OS-1, file "1/20/77–1/20/81"; "Remarks at the Congressional Space Medal of Honor Awards Ceremony," October 1, 1978.

107. "1979 Spring Planning Review, National Aeronautics and Space Administration, Issue #2: Solar Power Satellite Concept." Jimmy Carter Library, Domestic Policy Staff, Stern Files, Box 42, file "NASA: Budget, FY 1979, Spring Planning Review."

108. "NRC Study Recommends against Satellite Power System: Estimated $3 Trillion Price Tag Primary Reason," *Defense Daily* (July 7, 1981), 26.

109. Rockwell International, Space Division, *Industries in Space to Benefit Mankind: A View Over the Next 30 Years* (Washington: DC: NASA, SD 77-AP-0094), 2–4.

110. "An Era of Constraints," Stuart Eizenstat remarks at Women's National Democratic Club, January 4, 1979. Jimmy Carter Library, Domestic Policy Staff, Special Projects Stern, Box 87, file "Speeches: [Domestic Policy Staff]."

111. Sean Wilentz, *The Age of Reagan: A History, 1974–2008* (New York: HarperCollins, 2008), 128.

112. Jerry Adler et al., "In Space to Stay," *Newsweek* (April 27, 1981), 28.

113. Rick Gore, "When the Space Shuttle Finally Flies," *National Geographic* (March 1981), 320, 347.

114. Richard S. Lewis, *The Voyages of Columbia: The First True Spaceship* (New York: Columbia University Press, 1984), 98, 100.

115. George Torres, *Space Shuttle: A Quantum Leap* (Novato, CA: Presidio Press, 1986), 121.

116. Philip M. Boffey, "President Backs U.S. Space Station as Next Key Goal," *New York Times* (January 26, 1984), A1.

117. Hans Mark, *The Space Station: A Personal Journey* (Durham, NC: Duke University Press, 1987); Howard E. McCurdy, "The Space Station Decision: Politics, Bureaucracy, and the Making of Public Policy," in Martin J. Collins and Sylvia D. Fries, eds., *A Spacefaring Nation: Perspectives on American Space History and Policy* (Washington, DC: Smithsonian Institution Press, 1991), 9–28.

118. Philip M. Boffy, "President Backs U.S. Space Station as Next Key Goal," *New York Times* (January 26, 1984), B7.

119. "Amusement Parks of the Mind," *Sky* (August 1997), 113–118.

120. *New York Times* (May 17, 1987), 390.

121. *Space Camp*, ABC Motion Pictures, 1986.

122. "Visitor Exit Poll Summary, Where Next, Columbus? Exhibition," August 1993, December 1994, and August 1995. Department of Space History, NASM.

123. Young Astronauts Council, *Young Astronaut Program Fact Sheet*, 1988.

124. Ronald Reagan letter to Young Astronauts on the foundation's first anniversary, October 17, 1985.

125. Charles H. Holbrow et al. eds., *Space Colonization: Technology and the Liberal Arts* (Geneva, NY: American Institute of Physics, 1986), Foreword.

126. "Remarks of the Vice President Announcing the Winner of the Teacher in Space Project," July 19, 1985.

127. *Teacher in Space* (Washington, DC: NASA, 1985), 9.

128. *We Deliver* (Washington, DC: NASA, 1983), 2.

129. James Beggs speech before the National Press Club, November 5, 1982, NHO, file 004282; David Osbourne, "Business in Space," *The Atlantic Monthly* (May 1985), 52, 57; Office of Technology

Assessment, *Civilian Space Policy and Applications* (Washington, DC: U.S. Government Printing Office, 1982), 6–8.

130. "National Security Decision Directive Number 42, 'National Space Policy,' July 4, 1982," *Exploring the Unknown*, 1995, 591.

131. *Pioneering the Space Frontier: The Report of the National Commission on Space* (New York: Bantam Books, May 1986), 1–14.

132. Daniel Cohen, *Heroes of the Challenger* (New York: Pocket Books, 1986), 4.

133. *Regaining the Competitive Edge*, Howard Smith narrator (NASA, 1987).

134. Presidential Commission on the Space Shuttle Challenger Accident, *Report to the President*, June 9, 1986.

135. Jon D. Miller, *TheImpact of the Challenger Accident on Public Attitudes toward the Space Program: A Report to the National Science Foundation* (January 25, 1987), i; Louis Harris, "Despite Doubts about Shuttle Program, Solid Majorities Continue to Support Funding," *The Harris Survey* (March 6, 1986).

136. James W. McCulla (NASA public affairs officer) memo to NASA Administrator, "The Status of Public Opinion about NASA and the Space Program," June 3, 1986. NHO, file "James Fletcher, Public Affairs."

137. Presidential Commission on the Space Shuttle Challenger Accident, *Report to the President*, June 9, 1986 Preface.

138. Business-Higher Education Forum, *Space: America's New Competitive Frontier* (Washington, DC: NHO, April 1986).

139. Ronald Reagan, "Address to the Nation on the Explosion of the Space Shuttle Challenger," January 28, 1986.

140. Alex Roland, "The Shuttle: Triumph or Turkey," *Discover* (November 1985), 29–49.

141. "Can the Next President Win the Space Race?," *Nova* (October 11, 1988). LOC Film and Television Archives.

142. "From Disaster to Recovery," *CBS 48 Hours* (April 27, 1987). LOC Film and Television Archives.

143. *Voyage through the Universe: Starbound* (Alexandria, VA: Time-Life Books, 1989); *Voyage through the Universe: Spacefarers* (Alexandria, VA: Time-Life Books, 1989). Also: William K. Hartmann et al., *Out of the Cradle: Exploring the Frontiers beyond Earth* (New York: Workman Publishing, 1984); Thomas R. McDonough, *Space: The Next Twenty-Five Years* (New York: John Wiley, 1987); Frank White, *The Overview Effect* (Boston, MA: Houghton Mifflin, 1987); Harry L. Shipman, *Humans in Space: 21st Century Frontier* (New York: Plenum Press, 1989); Nicholas Booth, *Space: The Next 100 Years* (London: Mitchell Beazley, 1990); Richard S. Lewis, *Space in the 21st Century* (New York: Columbia University Press, 1990); George Henry Elias, *Breakout into Space: Mission for a Generation, an Argument for the Settlement of Space* (New York: William Morrow, 1990).

144. *Planetary Explorer: The Emigrant Trail, December 2038*, Marketing Communications, General Dynamics Space Division, 1988.

145. Mercer Cross, "Space Explorers Soon May Live in Self-Sustaining Biospheres," *National Geographic News Service* (April 16, 1986); Space Biosphere Ventures, *Biosphere II: A Project to Create a Biosphere*, 1987; "Down to Earth Space Station," *National Geographic World* (November 1988), 19; John Allen and Mark Nelson, *Space Biospheres* (Malabar, FL: Orbit Book, 1987).

146. Saul Pett, "Principles for Space Colonies Sought," *Richmond Times-Dispatch* (February 1, 1987); Justice William J. Brennan, Jr., "Space Settlements and the Law," Address to the American Law Institute, May 21, 1987. NASM Archives, file OS-205000–15.

147. George E. Brown, Jr., "The Space Settlement Act of 1988," *Congressional Record*, March 22, 1988.

148. Dr. Sally K. Ride, *Leadership and America's Future in Space* (Washington, DC: NASA, August 1987), 5, 58.

149. "Planning for the Future: NASA's New Office of Exploration," *Astronomy* (January 1988), 32, 36; Office of Exploration, *Beyond Earth's Boundaries: Human Exploration of the Solar System in the 21st Century* (Washington, DC: NASA, 1988), 40–43.

150. *1989 Long-Range Plan* (Washington, DC: NASA, 1989).

151. "Remarks on the 20th Anniversary of the *Apollo 11* Moon Landing," July 20, 1989, *Public Papers of the Presidents, Administration of George Bush, 1989* (Washington, DC: Office of Federal Register, 1990), 992.

152. Republican Party Platform of 1988. The American Presidency Project. http://www.presidency.ucsb.edu.

153. Presidents of the National Academy of Science and National Academy of Engineering joint letter to President-elect George Bush, December 12, 1988. George H. W. Bush Presidential Library, WHORM, OA/ID 19078, Box 1, file 003372.

154. Senators Richard Shelby and Jake Garn letter to President Bush, July 19, 1989. George H. W. Bush Presidential Library, WHORM, OA/ID 19078, Box 1, file 58179.

155. "Exploration Fact Sheet," NASA. George H. W. Bush Presidential Library, White House Press Office, Judy Smith Files, OA/ID 13273, Box 15.

Conclusion: The End of American Frontier Nationality?

1. Synthesis Group on America's Space Exploration Initiative, *America at the Threshold* (Washington, DC: US Government Printing House, 1991), iv.

2. "Remarks on Greeting the Crew of the Space Shuttle Discovery," March 24, 1989.

3. John Burgess, "Can the US Get Things Right Anymore?," *Washington Post* (July 3, 1990), C1.

4. "A Valuable Space Jalopy," *New York Times* (August 27, 1997), A22.
5. Seth Schiesel, "New Disney Vision Making the Future a Thing of the Past," *New York Times* (February 23, 1997), A1.
6. Daniel S. Goldin and Alex Roland, "Colonizing Space: What is our Goal?," Werhner von Braun Memorial Lecture, April 29, 1993 (Washington, DC: Smithsonian National Air and Space Museum, 1993), 2.
7. Marcia Smith, "A Congressional Perspective on the Space Exploration Initiative," E. Brian Pritchard ed., *Mars: Past, Present, and Future* (Washington, DC: American Institute of Aeronautics and Astronautics, 1992), 112.
8. John Noble Wilford, "Discovering Columbus," *New York Times Magazine* (August 11, 1991), 49.
9. Patricia Nelson Limerick, *The Legacy of Conquest: The Unbroken Past of the American West* (NY: Norton, 1987).
10. Peter Carlson, "Is NASA Necessary?," *Washington Post Magazine* (May 30, 1993), 22.
11. Francis Fukuyama, *The End of History and the Last Man* (NY: Free Press, 1992).
12. Ray Villard, "Earth in Danger!!," *ODYSSEY* (October 1992), 27.
13. William J. Broad, "Earth is Target for Space Rocks at Higher Rate than Thought," *New York Times* (January 7, 1997), C1.
14. William J. Broad, "US Will Deploy Its Spy Satellites on Nature Mission," *New York Times* (November 27, 1995), A1.
15. *Biosphere II: A Project to Create a Biosphere* (Space Biosphere Ventures, 1987); William J. Broad, "Paradise Lost: Biosphere Retooled as Atmospheric Nightmare," *New York Times* (November 19, 1996), C1.
16. National Research Council, *Science and Stewardship in the Antarctic* (Washington, DC, 1993), 1–2.
17. David Sanger and Richard Stevenson, "Bush Backs Goal of Flight to Moon to Establish Base," *New York Times* (January 15, 2004), A1.

Index

CPSIA information can be obtained
at www.ICGtesting.com
Printed in the USA
LVOW13*1428061216
516054LV00011B/345/P